T0173582

Cambridge International AS & A Level Mathematics

Mechanics

STUDENT'S BOOK

Tom Andrews, Michael Kent
Series Editor: Dr Adam Boddison

William Collins' dream of knowledge for all began with the publication of his first book in 1819.

A self-educated mill worker, he not only enriched millions of lives, but also founded a flourishing publishing house. Today, staying true to this spirit, Collins books are packed with inspiration, innovation and practical expertise. They place you at the centre of a world of possibility and give you exactly what you need to explore it.

Collins. Freedom to teach.

Published by Collins
An imprint of HarperCollins*Publishers*
The News Building
1 London Bridge Street
London
SE1 9GF

Browse the complete Collins catalogue at
www.collins.co.uk

British Library Cataloguing in Publication Data
A catalogue record for this publication is available from the British Library.

Commissioning editor: Jennifer Hall
In house editor: Lara McMurray
Authors: Tom Andrews/Michael Kent
Series editor: Dr Adam Boddison
Development editor: Tim Major
Project manager: Emily Hooton
Copyeditor: Jan Schubert
Proofreaders: Julie Bond/Joan Miller
Answer checkers: Julie Bond/David Hemsley
Cover designer: Gordon MacGilp
Cover illustrator: Maria Herbert-Liew
Typesetter: Jouve India Private Ltd
Illustrators: Jouve India Private Ltd/Ken Vail Graphic Design
Production controller: Sarah Burke
Printed and bound by Ashford Colour Press Ltd
®IGCSE is a registered trademark

This book is produced from independently certified FSC™ paper to ensure responsible forest management.

For more information visit:
www.harpercollins.co.uk/green

Acknowledgements

The publishers wish to thank Cambridge Assessment International Education for permission to reproduce questions from past IGCSE® Mathematics and AS & A Level Mathematics and Further Mathematics papers. Cambridge Assessment International Education bears no responsibility for the example answers to questions taken from its past papers. These have been written by the authors. Exam-style questions and sample answers have been written by the authors.

The publishers wish to thank the following for permission to reproduce photographs. Every effort has been made to trace copyright holders and to obtain their permission for the use of copyright material. The publishers will gladly receive any information enabling them to rectify any error or omission at the first opportunity.

pvi Cultura Creative (RF)/Alamy Stock Photo, p1 Cultura Creative (RF)/Alamy Stock Photo, p36 Henry Lederer/Taxi/Getty Images, p75t Vadim Sadovski/Shutterstock, p75b Denis Belitsky/ Shutterstock, p79 O2creationz/Shutterstock, p84 LeStudio/Shutterstock, p88 3Dsculptor/ Shutterstock, p97 Jeremy Walker/The Image Bank/ Getty Images, p144 Nuno Andre/Shutterstock.

HarperCollins*Publishers*
Macken House
39/40 Mayor Street Upper
Dublin 1
DO1 C9W8
Ireland

Full worked solutions for all exercises, exam-style questions and past paper questions in this book available to teachers by emailing international.schools@harpercollins.co.uk and stating the book title.

CONTENTS

Full worked solutions for all exercises, exam-style questions and past paper questions in this book available to teachers by emailing international.schools@harpercollins.co.uk and stating the book title.

INTRODUCTION

This book is part of a series of nine books designed to cover the content of the Cambridge International AS and A Level Mathematics and Further Mathematics syllabuses. The chapters within each book have been written to mirror the syllabus, with a focus on exploring how the mathematics is relevant to a range of different careers or to further study. This theme of *Mathematics in life and work* runs throughout the series, with regular opportunities to deepen your knowledge through group discussion and exploring real-world contexts.

Within each chapter, examples are used to introduce important concepts and practice questions are provided to help you to achieve mastery. Developing skills in modelling, problem solving and mathematical communication can significantly strengthen overall mathematical ability. The practice questions in every chapter have been written with this in mind and selected questions include symbols to indicate which of these underlying skills are being developed. Exam-style questions are included at the end of each chapter and a bank of practice questions, including real Cambridge past exam questions, is included at the end of the book.

A range of other features throughout the series will help to optimise your learning. These include:

> **key information boxes** – highlighting important learning points or key formulae

> **commentary boxes** – tackling potential misconceptions and strengthening understanding through probing questions

> **stop and think** – encouraging independent thinking and developing reflective practice.

Key mathematical terminology is listed at the beginning of each chapter and a glossary is provided at the end of each book. Similarly, a summary of key points and key formulae is provided at the end of each chapter. Where appropriate, alternative solutions are included within the worked solutions to encourage you to consider different approaches to solving problems.

This Mechanics book will introduce the concepts required to model real-life physical situations. You will learn how to apply Newton's three laws of motion alongside the principles of conservation of energy and conservation of momentum. Together, these principles and concepts can be used to model what happens when a car rolls down a ramp, when a ball is thrown in the air or when two objects collide. Mechanics is an area of mathematics that is directly applicable to a broad range of careers, including engineering, sports science and collision investigation.

Mechanics provides direct applications for a range of pure mathematics topics, including trigonometry and inequalities. In particular, the application of differentiation and integration in the context of acceleration, velocity and displacement is explored in this module, demonstrating the value of calculus in solving real-life problems.

MODELLING IN MECHANICS

When modelling mechanics problems, it is common to use diagrams as a visual aid, which may include a range of measures such as force, acceleration, velocity, speed and distance. The convention is to use different types of arrow to represent different measures, as shown below.

———→ **Force**

———≫ **Acceleration**

———⇢ **Velocity / Speed**

←——→ **Distance**

Particular 'modelling words' have specific meanings in the context of mechanics problems. The table below includes some of the modelling words you can expect to encounter.

Modelling word	Assumption
Light	The object has no mass
Smooth	There is no friction
Rough	There is friction
Inextensible / Inelastic	The object cannot be stretched or squashed
Uniform	The same throughout (for example, uniform velocity)
Particle	A single point representing an object
Rigid	The object cannot bend
String	A line with no thickness
Rod	A rigid straight line with no thickness

FEATURES TO HELP YOU LEARN

Mathematics in life and work

Each chapter starts with real-life applications of the mathematics you are learning in the chapter to a range of careers. This theme is picked up in group discussion activities throughout the chapter.

Learning objectives

A summary of the concepts, ideas and techniques that you will meet in the chapter.

Language of mathematics

Discover the key mathematical terminology you will meet in this chapter. Throughout the chapter, key words are written in bold. The words are defined in the glossary at the back of the book.

Prerequisite knowledge

See what mathematics you should know before you start the chapter, with some practice questions to check your understanding.

Explanations and examples

Each section begins with an explanation and one or more worked examples, with commentary, where appropriate, to help you follow it. Some show alternative solutions to get you thinking about different approaches to a problem.

---

1 FORCES AND EQUILIBRIUM

Mathematics in life and work

Whenever you push a trolley or pull a rope, you are applying a force. This chapter will introduce you to how pure mathematics can be applied to physical concepts; in this case, forces in equilibrium. The models used in this chapter are simplified versions of real-life situations used in industry – for example:

> If you were a civil engineer designing a bridge, you would need to ensure that your bridge was strong enough and sturdy enough to support pedestrians or vehicles that would use the bridge.

> If you were a mechanical engineer designing car tyres, you would need to ensure that they were not so smooth as to make driving dangerous or so rough as to prevent motion.

> If you were an architect designing a ramp to be used by people in wheelchairs, you would consider the angle of the slope and the material used, to make sure that the wheelchair could be pushed up or down the slope easily.

In this chapter, you will consider the thought processes required when designing a ramp.

LEARNING OBJECTIVES

You will learn how to:

> identify the forces acting in a given situation
> understand the vector nature of a force, and find and use the components and resultants
> use the principle that, when a particle is in equilibrium, the vector sum of the forces acting is zero, or, equivalently, that the sum of the components in any direction is zero
> understand that a contact force between two surfaces can be represented by two components: the normal component and the frictional component
> use the model of a 'smooth' contact and understand the limitations of this model
> understand the concepts of limiting friction and limiting equilibrium, recall the definition of coefficient of friction and use the relationship $F = \mu R$ or $F \leqslant \mu R$, as appropriate
> use Newton's third law.

LANGUAGE OF MATHEMATICS

Key words and phrases you will meet in this chapter:

coefficient of friction, component, equilibrium, force, friction, limiting equilibrium, magnitude, normal reaction, resolve, resultant force, rough, smooth, tension, thrust, vector, weight

---

Example 5

A peg bag of weight 20 N is suspended from two strings, one at 11° to the horizontal and the other at 7° to the horizontal. Find the magnitude of the tension in each string.

Because the strings are at different angles to each other, and because you are not told that it is one single string, you need to label the tensions differently.

Solution

Start by representing the information in a diagram. Let the tensions be T_1 and T_2.

By resolving horizontally and vertically, you can obtain two simultaneous equations in T_1 and T_2.

Resolve horizontally.

vi

Colour-coded questions

Questions are colour-coded (green, blue and red) to show you how difficult they are. Exercises start with more accessible (green) questions and then progress through intermediate (blue) questions to more challenging (red) questions.

1 A wardrobe is at rest on a rough slope inclined at θ to the horizontal, where $\sin\theta = \frac{5}{13}$. The wardrobe is on the point of slipping down the slope. Find the coefficient of friction between the wardrobe and the slope.

2 A particle is acted upon by three horizontal forces, $P\,N$, $Q\,N$ and $25\,N$, as shown in the diagram. The $Q\,N$ force acts due east and the $25\,N$ force acts due south. The $P\,N$ force acts on a bearing of $325°$. Given that the particle is in equilibrium, find:

a the value of P

b the value of Q.

Question-type indicators

The key concepts of problem solving, communication and mathematical modelling underpin your A level Mathematics course. You will meet them in your learning throughout this book and they underpin the exercises and exam-style questions. We have labelled selected questions that are especially suited to developing one or more of these key skills with these icons:

(PS) **Problem solving** – mathematics is fundamentally problem solving and representing systems and models in different ways. These include: algebra, geometrical techniques, calculus, mechanical models and statistical methods. This icon indicates questions designed to develop your problem-solving skills. You will need to think carefully about what knowledge, skills and techniques you need to apply to the problem to solve it efficiently.

These questions may require you to:

> use a multi-step strategy

> choose the most efficient method, or bring in mathematics from elsewhere in the curriculum

> look for anomalies in solutions

> generalise solutions to problems.

(C) **Communication** – communication of steps in mathematical proof and problem solving needs to be clear and structured, using algebra and mathematical notation, so that others can follow your line of reasoning. This icon indicates questions designed to develop your mathematical communication skills. You will need to structure your solution clearly to show your reasoning and you may be asked to justify your conclusions.

These questions may require you to:

> use mathematics to demonstrate a line of argument

> make use of mathematical notation in your solution

> follow mathematical conventions to present your solution clearly

> justify why you have reached a conclusion.

Mathematical modelling – a variety of mathematical content areas and techniques may be needed to turn a real-world situation into something that can be interpreted through mathematics. This icon indicates questions designed to develop your mathematical modelling skills. You will need to think carefully about what assumptions you need to make to model the problem and how you can interpret the results to give predictions and information about the real world.

These questions may require you to:

> construct a mathematical model of a real-life situation, using a variety of techniques and mathematical concepts

> use your model to make predictions about the behaviour of mathematical systems

> make assumptions to simplify and solve a complex problem.

Key information

These boxes highlight information that you need to pay attention to and learn, such as key formulae and learning points.

KEY INFORMATION

The unit of force is the newton (N).

Stop and think

These boxes present you

Stop and think — How might the reaction and weight forces be related if the vase was on a table inclined on a table at 1° to the horizontal? Would it matter whether the table was smooth or rough? What would happen if the angle was increased?

with probing questions and problems to help you to reflect on what you have been learning. They challenge you to think more widely and deeply about the mathematical concepts, tackle misconceptions and, in some cases, generalise beyond the syllabus. They can be a starting point for class discussions or independent research. You will need to think carefully about the question and come up with your own solution.

Mathematics in life and work – Group discussions give you the chance to apply the skills you have learned to a model of a real-life maths problem from a career that uses maths. Your focus is on applying and practising the concepts and coming up with your own solutions, as you would in the workplace. These tasks can be used for class discussions, group work or as an independent challenge.

Summary of key points

At the end of each chapter, there is a summary of key formulae and learning points.

Exam-style questions

Practise what you have learnt throughout the chapter with questions, written in examination style by our authors, progressing in order of difficulty.

The last **Mathematics in life and work** question draws together the skills that you have gained in this chapter and applies them to a simplified real-life scenario.

At the end of the book, test your mastery of what you have learned in the **Summary review** section. Practise the basic skills with some Cambridge IGCSE questions, and then go on to try carefully selected questions from Cambridge International A Level past exam papers. Extension questions, written by our authors, give you the opportunity to challenge yourself and will prepare you for more advanced study.

Warm-up Questions		Extension Questions
Three Cambridge IGCSE® past paper questions based on prerequisite skills and concepts that are relevant to the main content of this book.	Selected past paper exam questions and exam-style questions on the topics covered in this syllabus component.	Extension questions that give you the opportunity to challenge yourself and prepare you for more advanced study.

1 FORCES AND EQUILIBRIUM

Mathematics in life and work

Whenever you push a trolley or pull a rope, you are applying a force. This chapter will introduce you to how pure mathematics can be applied to physical concepts; in this case, forces in equilibrium. The models used in this chapter are simplified versions of real-life situations used in industry – for example:

> If you were a civil engineer designing a bridge, you would need to ensure that your bridge was strong enough and sturdy enough to support pedestrians or vehicles that would use the bridge.

> If you were a mechanical engineer designing car tyres, you would need to ensure that they were not so smooth as to make driving dangerous or so rough as to prevent motion.

> If you were an architect designing a ramp to be used by people in wheelchairs, you would consider the angle of the slope and the material used, to make sure that the wheelchair could be pushed up or down the slope easily.

In this chapter, you will consider the thought processes required when designing a ramp.

LEARNING OBJECTIVES

You will learn how to:

> identify the forces acting in a given situation

> understand the vector nature of a force, and find and use the components and resultants

> use the principle that, when a particle is in equilibrium, the vector sum of the forces acting is zero, or, equivalently, that the sum of the components in any direction is zero

> understand that a contact force between two surfaces can be represented by two components: the normal component and the frictional component

> use the model of a 'smooth' contact and understand the limitations of this model

> understand the concepts of limiting friction and limiting equilibrium, recall the definition of coefficient of friction and use the relationship $F = \mu R$ or $F \leqslant \mu R$, as appropriate

> use Newton's third law.

LANGUAGE OF MATHEMATICS

Key words and phrases you will meet in this chapter:

coefficient of friction, component, equilibrium, force, friction, limiting equilibrium, magnitude, normal reaction, resolve, resultant force, rough, smooth, tension, thrust, vector, weight

PREREQUISITE KNOWLEDGE

You should already know how to:

- solve simultaneous linear equations in two unknowns
- apply Pythagoras' theorem and the sine, cosine and tangent ratios for acute angles to the calculation of a side or of an angle of a right-angled triangle
- solve problems using the sine and cosine rules for any triangle.

You should be able to complete the following questions correctly:

1 Solve the simultaneous equations.

$5x + 4y = 46$ ①

$2x = 3y$ ②

2 A right-angled triangle UVW has angle $UVW = 90°$, $VW = 14\,cm$ and angle $VUW = 58°$. Find the length of UW.

3 A right-angled triangle EFG has angle $EFG = 90°$, $EG = 19\,m$ and $FG = 15\,m$. Find the size of the angle EGF.

4 In the diagram, ABD is a triangle and ABC is a straight line. The angle CBD is $58°$. $AB = 19\,cm$ and $BD = 13\,cm$.

   a Use the cosine rule to find the length of the side AD.

   b Use the sine rule to find the size of the angle BAD.

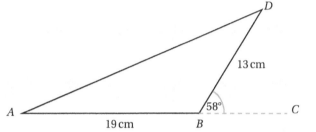

1.1 Forces

If an object experiences a push or a pull, it is said to be acted upon by a **force**. Force is measured in newtons (N). For example, in the diagram below, there are four forces acting on an object. The force T is acting horizontally to the right, the force F is acting horizontally to the left, the force R is acting vertically upwards and the force W is acting vertically downwards.

Forces are usually drawn coming out of the particle they are acting upon, with a line and a black-headed arrow showing the direction of action of the force.

A force is a **vector** quantity, which means that it has both a **magnitude** (size) and direction. Hence the directions of T, F, R and W are important. For example, if $T = 18\,N$ and $F = 14\,N$, then there would be an overall net force (known as the **resultant force**) of $4\,N$ acting to the right.

If two equal and opposite forces act upon a particle, such as if $T = 18\,N$ and $F = 18\,N$ and if $R = W$, then the forces cancel each other out and the resultant force is zero. When the resultant force is zero, this is known as **equilibrium**.

> **KEY INFORMATION**
>
> The unit of force is the newton (N).

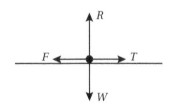

> **KEY INFORMATION**
>
> A force is a vector quantity. It has a magnitude and a direction.

If two forces act in opposite directions, they have opposite signs (positive and negative). If a force acts vertically, then it has no horizontal **component** and if a force acts horizontally, it has no vertical component.

Newton's third law

Newtons' laws are covered in more depth in **Chapter 4 Newton's laws of motion**. Much of the material in this chapter, however, requires an understanding of Newton's second and third laws.

Newton's second law states that $F = ma$ (resultant force is proportional to mass × acceleration). Note that in this chapter objects will be stationary or travelling at a steady speed, and hence the acceleration and resultant force will both be zero.

Newton's third law states that every action has an equal and opposite reaction. From observation, you know that your **weight** is balanced by a force from the ground that prevents you from falling into it. This is true for any object on a stable, solid surface, such as a vase on a table or a car on a road.

Types of force

The five types of force you need to know are weight, normal reaction, tension, thrust and friction.

Weight

The weight of an object is a force caused by the gravitational acceleration experienced by the object and acts vertically downwards towards the centre of the Earth.

In this diagram, the weight is labelled as W. Because weight acts vertically downwards, the arrow in the diagram also points vertically downwards.

Normal reaction

If an object is in contact with a surface, the object will experience a reaction force perpendicular to the surface. This is often called the **normal reaction**, where 'normal' means 'perpendicular'. In a diagram, the normal reaction is usually labelled R.

Hence, in a diagram with a horizontal surface, the reaction force acts vertically because it always acts perpendicular to the surface.

However, in a diagram with a slope, the reaction force will act at an angle to the horizontal, but perpendicular to the slope.

The reaction force is the one often described as the force exerted on the particle by the ground since it is equal and opposite to the weight of the particle when the ground is horizontal. This is in agreement with Newton's third law, as indicated above, which states that every action has an equal and opposite reaction.

> **KEY INFORMATION**
>
> The sum of all the forces is called the resultant force.
>
> If the resultant force is equal to zero, the object is in equilibrium.

W

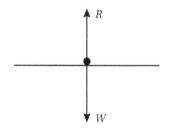

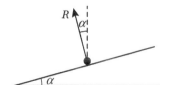

Tension

When an object is pulled by or suspended from a string or rope, the string will become taut and the object will experience the **tension** in the string.

In a diagram, tension is usually labelled T and acts away from the object.

Thrust (compression)

Consider a ball held in mid-air. If the ball is suspended from a string so that the string is above the ball and taut, the force in the string is tension. On the other hand, if the ball is held up on a wooden rod, then the force in the wooden rod is **thrust** (or compression). When an object is pushed rather than pulled, the object will experience a thrust force.

In this diagram, the particle is held up by a rod rather than being suspended from a string, so the T stands for thrust. In the case of thrust, since it is a pushing force, the force appears to be going *into* the particle rather than coming out of it.

In summary, tension always acts *away* from an object whereas thrust always acts *towards* the object.

Friction

When you brush your hand across a table, you experience a resistive force (or a resistance to motion). This force is called **friction**. In a diagram, friction is usually labelled F. In the diagram to the right, F has been used to represent

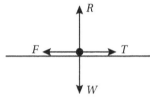

the friction when a particle is pulled by a string along a **rough** table. Note that if a surface is not rough, it is described as **smooth**.

> **KEY INFORMATION**
> Friction (F) is a force caused by electromagnetic interactions between surfaces and it opposes motion. When a pair of surfaces is described as smooth they are assumed to have no electromagnetic interactions between them and $F = 0$. When a pair of surfaces is described as rough they are assumed to have electromagnetic interactions between them and $0 < F \leqslant \mu R$. This is explored further in **Section 1.4**.

Example 1

A vase is at rest on a horizontal table. The weight of the vase is 25 N.

Find the force exerted on the vase by the table.

Solution

Start by drawing a diagram to represent the information. Model the vase as a particle. Draw a 25 N force acting vertically downwards for the weight of the vase. The force exerted on the vase is the reaction force, which acts perpendicular to the surface (i.e. vertically upwards).

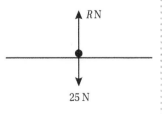

Since both forces act vertically and there is no acceleration, the two forces are balanced, in accordance with Newton's third law.

Hence $R = 25\,\text{N}$.

Stop and think How might the reaction and weight forces be related if the vase was on a table inclined on a table at 1° to the horizontal? Would it matter whether the table was smooth or rough? What would happen if the angle was increased?

Exercise 1.1A

1 Draw diagrams for the following.

   a A particle is at rest on a horizontal surface.

   b A suitcase is pulled across a rough horizontal table by a horizontal rope.

   c A particle is pushed across a smooth horizontal table.

2 A particle experiences four horizontal forces, A, B, C and D, as shown in the diagram.

If the particle is stationary, what can you say about the forces?

3 A toy train is pulled along a horizontal table at a steady speed by a taut string. The frictional force is given by X. The weight of the train is $4\,\text{N}$. Label the missing forces in the diagram.

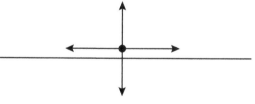

4 A box is pushed by a thrust force at a steady speed along a rough horizontal table. The frictional force is $12\,\text{N}$. The reaction force exerted on the box by the table is $58\,\text{N}$.

   a Draw a force diagram to represent the four forces acting upon the box.

   b Calculate the magnitude of the thrust force.

   c Calculate the weight of the box.

1.2 Resolving forces

In **Section 1.1**, all the forces acted horizontally or vertically and you considered the forces acting in these two perpendicular directions.

The process of considering the forces acting in a particular direction is called **resolving**. In this section, the technique will be extended to a force acting at an angle to the horizontal or a particle on a slope. This means that the weight does not act parallel or perpendicular to the slope.

Resolving in a particular direction gives the resultant force in that direction.

Consider a force of F N acting at an angle of θ to the horizontal. You can construct a triangle of forces.

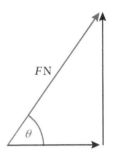

This is a right-angled triangle with the force F as its hypotenuse. You can therefore use trigonometry to split the force into its horizontal and vertical components.

The vertical component is the component opposite to θ.

Since $\sin \theta = \dfrac{\text{opposite (opp)}}{\text{hypotenuse (hyp)}} = \dfrac{O}{H}$, you can write $O = H \sin \theta$, and since F is the hypotenuse, the opposite is given by $F \sin \theta$.

Similarly, since $\cos \theta = \dfrac{\text{adjacent (adj)}}{\text{hypotenuse (hyp)}} = \dfrac{A}{H}$, you can write $A = H \cos \theta = F \cos \theta$.

Hence the component adjacent to the angle is given by $F \cos \theta$ and the component opposite the angle is given by $F \sin \theta$.

For example, in the diagram on the right, the 17 N force acts at angle of 50° to the horizontal. The horizontal component adjacent to the angle is given by $17 \cos 50° = 10.9$ N and the vertical component opposite the angle is given by $17 \sin 50° = 13.0$ N.

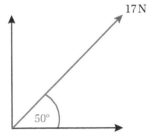

Example 2

A box of books is dragged along a rough horizontal floor by a rope at a steady speed. The rope makes an angle of 35° with the floor.

a Draw a force diagram.

b Resolve horizontally.

c Resolve vertically.

Solution

a The box of books is modelled as a particle, which is drawn as a black circle.

Assume that the box is being dragged from left to right along the floor.

There are four forces acting upon the box, each of which is represented as an arrow.

The weight of the box (W) acts vertically downwards.

The rope has a tension (T), and since the rope makes an angle of 35° with the floor, so does the tension force.

The reaction force (R) acts perpendicular to the surface. Since the surface (the floor) is horizontal, the reaction force is vertical.

The frictional force (F) acts to oppose the motion. Since the box is being dragged to the right, the friction acts to the left.

b Since the weight and reaction act vertically, they have no horizontal component. Hence there are only two forces to consider, the tension and the friction.

The tension acts at an angle of 35° to the horizontal, so has a component of $T\cos 35°$ horizontally.

The friction acts horizontally.

Because the box is being dragged at a steady speed, there is no acceleration. Hence the resultant force is zero.

Resolve horizontally to find the resultant force in the horizontal direction.

R($\rightarrow$)

$T\cos 35° - F = 0$

c Since the friction acts horizontally, it has no vertical component, and there are three forces to consider – the tension, weight and reaction.

The tension acts at an angle of 35° to the horizontal, so has a component of $T\sin 35°$ vertically.

The reaction and weight act vertically.

Because the box is not moving vertically, there is no acceleration. Hence the resultant force is zero.

Resolve vertically to find the resultant force in the vertical direction (you could choose up or down as positive).

R($\uparrow$)

$R + T\sin 35° - W = 0$

This agrees with Newton's third law.

KEY INFORMATION

Weight acts vertically downwards, the reaction acts perpendicular to the surface and friction acts to oppose the motion.

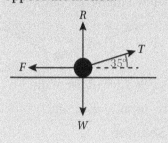

It is useful to consider which forces have a component in the direction in which you are resolving and which do not before you begin.

This notation, R($\rightarrow$), is used to indicate that you are resolving in the direction of the arrow.

Note that when forces are in equilibrium and there is no resultant force, you may put the balanced forces equal to each other (as shown in **Section 1.1**) or you may put the sum of the forces equal to zero (as shown here).

This component could also be written as $T\cos 55°$.

Example 3

A wetsuit of weight W N hanging from a washing line causes the line to make angles of 22° and 33° with the horizontal, as shown in the diagram.

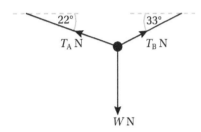

a Resolve horizontally.

b Resolve vertically.

Solution

a Because the angles are different, assume also that the tensions are different. Label these T_A and T_B. Also add in the weight of the wetsuit, acting vertically downwards.

Resolve horizontally to find the resultant force in the horizontal direction.

R($\rightarrow$)

$T_B \cos 33° - T_A \cos 22° = 0$

b Resolve vertically to find the resultant force in the vertical direction.

R($\uparrow$)

$T_A \sin 22° + T_B \sin 33° - W = 0$

KEY INFORMATION

For two identical strings, if the angles are the same then the magnitude of the tensions will also be the same. Similarly, if the tensions are the same, then the angles must be the same. If the angles are different, however, then the magnitude of the tensions cannot be assumed to be the same.

This technique applies to a force given at any angle. The components do not have to be horizontal and vertical. If a particle is on a surface which is itself at an angle, then it is common to resolve parallel and perpendicular to the slope, in which case the weight (and possibly other forces too) will need to be resolved into components. Note that the friction is always parallel to the surface and the reaction force is always perpendicular to the surface.

If a particle is *just* held in position by a force on a rough slope so that the particle is on the point of slipping down the slope, then

KEY INFORMATION

Always resolve parallel and perpendicular to the surface.

the frictional force acts up the slope. If the particle is *on the point of beginning to move* up a rough slope, then the frictional force acts down the slope.

Example 4

A toaster of weight W N is at rest on a rough slope inclined at θ to the horizontal. The toaster is acted upon by a horizontal force of X N and as a result is on the point of moving up the slope.

Resolve parallel and perpendicular to the slope.

Solution

Start by drawing a force diagram, representing the toaster as a particle.

There are four forces acting upon the toaster. All are measured in newtons (N).

The weight of the box (W) acts vertically downwards.

The horizontal force (X) must be acting to the right if it is to keep the particle on the slope.

The reaction force (R) acts perpendicular to the surface. Since the surface is at an angle, so is the reaction force.

The frictional force (F) acts to oppose the motion. Since the toaster is on the point of moving up the slope, the friction acts down the slope, parallel to the slope.

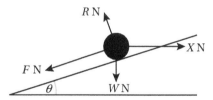

Start by resolving parallel to the slope. Since the reaction acts perpendicular to the slope, it has no parallel component. Therefore there are three forces to consider: X, the weight and the friction.

The friction acts parallel to the slope.

Since X is horizontal, the component parallel to the slope is equivalent to the side adjacent to the angle, as shown in the diagram below, and hence is given by $X\cos\theta$.

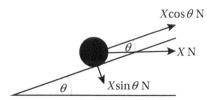

Since W is vertical, the component parallel to the slope is equivalent to the side opposite the angle, as shown in the diagram below, and hence is given by $W\sin\theta$.

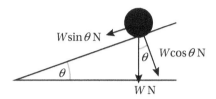

Resolve parallel to the slope.

R($\nearrow$)

$X\cos\theta - W\sin\theta - F = 0$

Now consider resolving perpendicular to the slope. Since the friction acts parallel to the slope, it has no perpendicular component. Again, there are three forces to consider: X, the weight, W, and the reaction, R.

The reaction acts perpendicular to the slope.

Since X is horizontal, the component perpendicular to the slope is equivalent to the side opposite the angle, as shown in the diagram below, and hence is given by $X\sin\theta$.

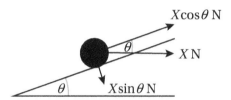

Since W is vertical, the component perpendicular to the slope is equivalent to the side adjacent to the angle, as shown in the diagram below, and hence is given by $W\cos\theta$.

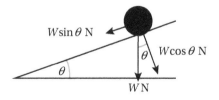

Resolve perpendicular to the slope.

R($\nwarrow$)

$R - X\sin\theta - W\cos\theta = 0$

> Alternatively, you could resolve down the slope instead of up the slope.

> You could have resolved towards the slope instead of away from the slope.

Exercise 1.2A

1 Draw diagrams for the following.

a A particle is at rest on a rough plane inclined at θ to the horizontal.

b A particle is pulled across a rough horizontal table by a string inclined at 60° to the horizontal.

c A particle is held at rest on a smooth plane inclined at 30° to the horizontal by a string parallel to the plane.

d A car drives up a rough slope inclined at θ to the horizontal.

e A particle is on the verge of moving up a rough slope (inclined at 26° to the horizontal) as a result of being pushed by a horizontal force P.

f A particle is just held in position on a rough slope (inclined at 26° to the horizontal) by a horizontal force P.

g A particle is suspended from a horizontal beam by two unequal strings at angles of 50° and 20° to the beam.

2 A peg bag of weight W N is suspended from a string such that both ends of the string are inclined at θ to the horizontal and the tensions in both ends are identical and given by T N, as shown in the diagram. The system is in equilibrium.

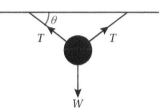

a Resolve vertically.

b Hence find an expression for T in terms of W and θ.

3 A toy bus of weight W N is pulled along a rough horizontal carpet at a constant speed by a string. The string makes an angle of 19° with the carpet, as shown in the diagram. The system is in equilibrium.

a Resolve parallel to the carpet.

b Resolve perpendicular to the carpet.

4 A particle of weight W N is suspended from two strings, A and B, as shown in the diagram. String A is at an angle of 25° to the horizontal and string B is at an angle of 65° to the horizontal. The system is in equilibrium.

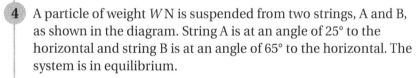
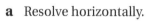

a Resolve horizontally.

b Resolve vertically.

5 Each diagram shows a system in equilibrium. For each diagram, Mary has been asked to resolve in a given direction. Explain what is wrong with Mary's answer in each case and give the correct answer.

a Resolve perpendicular to the slope.

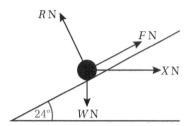

Mary's answer: $R = W\cos 24°$

b Resolve parallel to the slope.

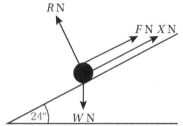

Mary's answer: $F + X\cos 24° = W\sin 24°$

c Resolve parallel to the slope.

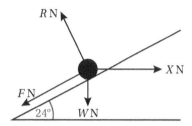

Mary's answer: $X\cos 24° = F + W\cos 24°$

d Resolve perpendicular to the slope.

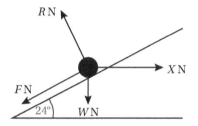

Mary's answer: $R = X\cos 24° + W\cos 24°$

e Resolve parallel to the slope.

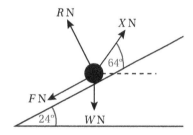

Mary's answer: $X\cos 64° = F + W\sin 24°$

6 A clock of weight W N is at rest on a rough slope inclined at angle $θ$. The clock experiences a normal reaction of R N and a frictional force of F N.

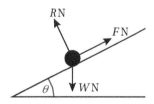

Show that $F = R\tan θ$.

7 A piano of weight W N is pushed up a rough slope at a steady speed by a horizontal force of X N. The slope is inclined at angle θ. The piano experiences a normal reaction of R N and a frictional force of F N.

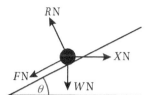

a Resolve parallel to the slope.

b Resolve perpendicular to the slope.

c Hence show that X can be written as $R\sin\theta + F\cos\theta$.

8 A freezer of weight 250 N is being pushed at a steady speed up a rough path inclined at an angle of 12° to the horizontal, as shown in the diagram. The pushing force, X N, is horizontal. Given that the magnitude of the frictional force is one-quarter the magnitude of the normal reaction force, show that X is approximately 120 N.

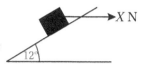

9 A box of weight W N is pulled along a rough horizontal road at a steady speed by a rope inclined at 30° to the horizontal, as shown in the diagram. R and F are the normal reaction and friction forces, respectively.

Show that W is given by the expression $\dfrac{3R + \sqrt{3}F}{3}$.

10 A block of weight W N is pulled at a steady speed up a rough plane inclined at an angle of 27° by a cable of tension T N inclined at an angle of 15° to the plane. The reaction and friction forces are given by R N and F N, respectively. Given that $F = \frac{1}{2}R$, show that $\dfrac{T}{W} = \dfrac{\cos 27 + 2\sin 27}{2\cos 15 + \sin 15}$.

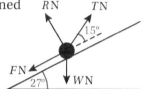

Mathematics in life and work: Group discussion

You are an architect and you have been asked to design a ramp for entry to a hospital. The entrance is 0.8 m above the ground. You have a rectangular plot measuring 5 m by 12 m on which to build the ramp. The purpose of the ramp is to allow access for wheelchairs.

1 What forces would act upon a wheelchair on the ramp? Which of these forces would have the greatest magnitude and which the smallest magnitude? How might this influence your choice of design?

2 What is the minimum width of a ramp that would enable a wheelchair to be pushed up or down it? Justify your answer.

3 If the ramp were 12 m long, what would be the angle of incline of the ramp? Comment on this angle in the context of pushing a wheelchair.

4 How might you reduce this angle of incline? What practical implications would you need to consider?

5 In a group, design the most effective ramp you can that would maximise the use of the space you have been allocated given the constraints.

If all the forces are in equilibrium, the sum in any direction will be zero. This means that it is possible to solve problems by setting up and solving equations.

Example 5

A peg bag of weight 20 N is suspended from two strings, one at 11° to the horizontal and the other at 7° to the horizontal. Find the magnitude of the tension in each string.

Because the strings are at different angles to each other, and because you are not told that it is one single string, you need to label the tensions differently.

Solution

Start by representing the information in a diagram. Let the tensions be T_1 and T_2.

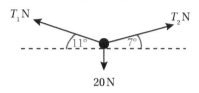

By resolving horizontally and vertically, you can obtain two simultaneous equations in T_1 and T_2.

Resolve horizontally.

R($\rightarrow$)

$$T_2 \cos 7° - T_1 \cos 11° = 0 \qquad ①$$

Resolve vertically.

R($\uparrow$)

$$T_1 \sin 11° + T_2 \sin 7° - 20 = 0 \qquad ②$$

Make T_1 the subject of equation ①.

This is an example of solving simultaneous equations by substitution.

$$T_1 = \frac{T_2 \cos 7°}{\cos 11°}$$

Substitute for T_1 in equation ②.

$$\frac{T_2 \cos 7°}{\cos 11°} \sin 11° + T_2 \sin 7° - 20 = 0$$

$$T_2 \left(\frac{\cos 7°}{\cos 11°} \sin 11° + \sin 7° \right) = 20$$

$$T_2 = \frac{20}{\frac{\cos 7°}{\cos 11°} \sin 11° + \sin 7°} = 63.5 \,\text{N}$$

$$T_1 = \frac{T_2 \cos 7°}{\cos 11°} = 63.5 \times \frac{\cos 7°}{\cos 11°} = 64.2 \,\text{N}$$

The tensions are 63.5 N and 64.2 N.

Alternative method

Draw the three forces as a triangle, 'tip to tail'.

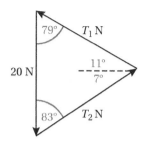

T_1 and T_2 are at 11° and 7°, respectively, to the horizontal, giving a total of 18°.

Using alternate angles or right-angled triangles, you can calculate the other angles as 79° and 83°.

Apply the sine rule.

$$\frac{20}{\sin 18°} = \frac{T_2}{\sin 79°}$$

$$T_2 = \frac{20\sin 79°}{\sin 18°} = 63.5 \text{ N}$$

Similarly, $T_1 = \frac{20\sin 83°}{\sin 18°} = 64.2$ N.

Another alternative method is to use Lami's theorem, although this is not a requirement of this course. Lami's theorem states that in a system with three coplanar, non-collinear forces, each force will be proportional to the sine of the angle between the other two forces. It is based upon the sine rule.

Lami's theorem can be written as

$$\frac{A}{\sin \alpha} = \frac{B}{\sin \beta} = \frac{C}{\sin \gamma}$$

where A, B and C are the forces (in newtons) and α, β and γ are the angles between the other two forces (for example, α is the angle between forces B and C).

For the peg bag example, the angle between T_1 and 20 N is 101° (90° + 11°), the angle between T_2 and 20 N is 97° (90° + 7°) and the angle between T_1 and T_2 is 162° (180° − (11° + 7°)).

To find T_1, substitute into the formula as follows.

$$\frac{T_1}{\sin 97°} = \frac{20}{\sin 162°}$$

Hence

$$T_1 = \frac{20\sin 97°}{\sin 162°} = 64.2\,\text{N}.$$

Similarly,

$$\frac{T_2}{\sin 101°} = \frac{20}{\sin 162°} = 63.5\,\text{N}.$$

The latter two methods can only be used for problems with three forces.

An angle is often given in the form of an inverse trigonometric function such as $\theta = \sin^{-1}\left(\frac{a}{b}\right)$. Rather than working out the angle, it is more elegant in this situation to use a right-angled triangle to find the other trigonometric ratios you require, either with a Pythagorean triple or in surd form. This will save you writing down the angle, which will usually be an irrational number and will lead to possible errors if rounded. Note that, in this context, the angles will all be acute, so their trigonometric ratios will be positive.

> If an angle is given in the form of an inverse trigonometric function, use Pythagoras' theorem and the tangent ratio to find any other trigonometric ratios.

For example, if $\theta = \sin^{-1}\left(\frac{5}{13}\right)$, sketch a right-angled triangle with an opposite of 5 and a hypotenuse of 13,

since $\sin\theta = \dfrac{\text{opposite (opp)}}{\text{hypotenuse (hyp)}} = \dfrac{O}{H}$.

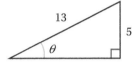

Using Pythagoras' theorem, the adjacent is given by

$$\sqrt{13^2 - 5^2} = \sqrt{144} = 12$$

Hence

$$\cos\theta = \frac{\text{adjacent (adj)}}{\text{hypotenuse (hyp)}} = \frac{A}{H} = \frac{12}{13}.$$

Similarly, if you were told that an angle was $\tan^{-1} 0.75$, you could rewrite 0.75 as $\frac{3}{4}$, so the opposite is 3 and the adjacent is 4

$$\left(\text{since } \tan\theta = \frac{\text{opposite (opp)}}{\text{hypotenuse (hyp)}} = \frac{O}{A}\right).$$

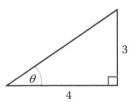

This is a 3, 4, 5 triangle, so the hypotenuse is 5.

Hence $\sin\theta = \frac{3}{5}$ and $\cos\theta = \frac{4}{5}$.

Example 6

A 25 N pulley is suspended from two identical light inextensible strings both at $\cos^{-1}\frac{8}{17}$ to the horizontal. Find the tension in each string.

Solution

Start by drawing the force diagram.

Let each tension be equal to T and call the angle θ. Note that θ can be placed at either the pulley or the other end of the strings, and the angles are equal since they are alternate.

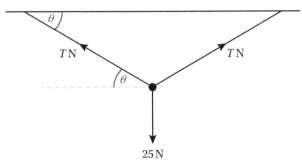

Resolving horizontally will not provide any additional information because the horizontal components of the tensions are equal and opposite.

Resolve vertically.

R($\uparrow$)

$T\sin\theta + T\sin\theta - 25 = 0$

$\qquad 2T\sin\theta - 25 = 0$

Hence

$$T = \frac{25}{2\sin\theta}$$

In this example, the angle θ has been given as $\cos^{-1}\frac{8}{17}$.

Sketch a right-angled triangle with an adjacent of 8 and a hypotenuse of 17, since $\cos\theta = \dfrac{\text{adjecent (adj)}}{\text{hypotenuse (hyp)}} = \dfrac{A}{H}$.

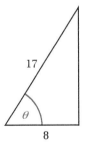

Using Pythagoras' theorem, the opposite is given by

$\sqrt{17^2 - 8^2} = \sqrt{225} = 15$.

Hence

$\sin\theta = \dfrac{\text{opposite (opp)}}{\text{hypotenuse (hyp)}} = \dfrac{O}{H} = \dfrac{15}{17}$.

Thus

$T = \dfrac{25}{2 \times \frac{15}{17}} = 14.2\,\text{N}$.

Exercise 1.2B

1 A 50 N chair is at rest on a slope inclined at 28° to the horizontal, as shown in the diagram. The chair is held in position by friction. Find the magnitude of:

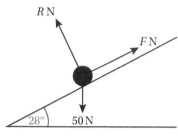

a the reaction force

b the friction force.

2 A W N particle is suspended from two identical strings.

Each string is inclined at an angle of $\sin^{-1}\left(\dfrac{12}{35}\right)$ to the horizontal.

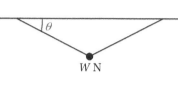

Given that the tension in each string is 32 N, find the value of W.

3 A particle of weight 38 N, suspended from a rope inclined at 20° to the vertical, is kept in position by a horizontal force X N, as shown in the diagram. Victrix and Colin have been asked to find the value of X, correct to 3 significant figures.

Victrix's solution is as follows.

R($\uparrow$) $T\cos 20° = 38$

$T = 38 \div \cos 20° = 40.439$ N

R($\rightarrow$) $X = T\sin 20° = 40.439 \times \sin 20° = 13.8$ N (3 s.f.)

Colin's solution is as follows.

R($\nearrow$)

$X\cos 20° - 38\sin 20° = 0$

$X\cos 20° = 38\sin 20°$

$X = 38\tan 20° = 13.8$ N

Compare and contrast the two solutions.

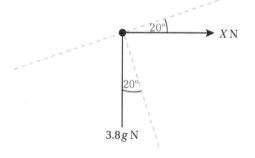

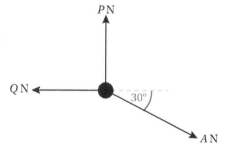

4 A particle at rest on a horizontal table experiences three horizontal forces, P, Q and A.

Forces P and Q are perpendicular. Force A acts at 30° to the line of action of Q, as shown in the diagram.

Q is 4 N larger than P.

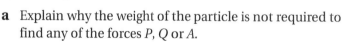

a Explain why the weight of the particle is not required to find any of the forces P, Q or A.

b Find the magnitude of A.

5 A case of weight 30 N is dragged along a rough horizontal floor by a cord inclined at $\sin^{-1} 0.2$ to the horizontal. The case experiences a frictional force of $\sqrt{6}$ N. Find the reaction exerted by the floor on the case.

6 A box of bricks weighs 120 N. The box is at rest on a rough plane inclined at an angle θ above the horizontal, where $\cos \theta = \left(\dfrac{40}{41}\right)$. A 20 N force acts on the particle up the line of greatest slope of the plane. Find the magnitude and direction of the frictional force.

7 The diagram shows a 15 N peg bag suspended from two unequal strings at 55° and 35° to the horizontal.

Find the tension in each string.

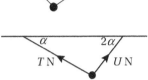

8 A pegbag of weight 20 N is suspended from a light cord such that one end of the cord, with a tension of T N, makes an angle of α with the horizontal and the other end, with a tension of U N, makes an angle of 2α with the horizontal, as shown in the diagram.

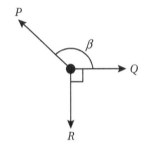

  a Show that $U = \dfrac{20\cos\alpha}{\cos 2\alpha \sin \alpha + \sin 2\alpha \cos \alpha}$.

  b Given that $\alpha = 30°$, show that $T = 10$.

9 Three horizontal forces, P, Q and R, act upon a particle, as shown in the diagram. Q and R are perpendicular. The angle between P and Q is obtuse and equal to β.

Show that $Q + R = P(\sin \beta - \cos \beta)$.

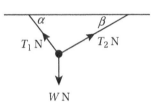

10 A pulley of weight W N is suspended from a string, as shown in the diagram.

One end of the string is inclined at an angle α, where $\cos \alpha = \dfrac{12}{37}$, with a tension of T_1 N.

The other end of the string is inclined at an angle β with a tension of T_2 N.

Given that T_1 is twice the magnitude of T_2, show that $W = \dfrac{T_2\left(70 + \sqrt{793}\right)}{37}$.

1.3 Adding forces

If two forces act upon a particle in different directions, then it is possible to add the forces together to find the resultant force.

If the forces are perpendicular, then you can use Pythagoras' theorem to find the magnitude and the tangent ratio to find the direction. If the forces are not perpendicular, then you will need to use the sine and cosine rules instead.

Note that an alternative name for the positive x-axis is O_x. If a force acts along the positive x-axis, it is said to act along O_x.

If more than two forces act upon a particle, then it is possible to apply the process as many times as necessary: an additional time for each additional force.

KEY INFORMATION

Forces can be added using the sine and cosine rules.

KEY INFORMATION

The positive x-axis can be written as O_x.

Example 7

A particle is acted upon by two forces, P and Q. P is a 12 N force acting along O_x, whereas Q is a 10 N force that acts at 40° to O_x, as shown in the diagram.

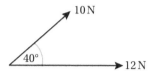

Find the magnitude and direction of the resultant force, both correct to 1 decimal place.

Solution

In order to use the triangle law, start by redrawing the forces tip to tail, as shown:

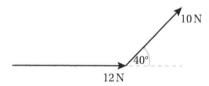

Because the 10 N force acts at 40° to O_x, the angle between the two forces is 140° since the two angles lie on a straight line. Thus the resultant force is the longest side in a triangle in which the two other sides have magnitude 12 and 10, and there is an angle of 140° between them, as shown. The resultant force can be calculated by using the cosine rule.

> Remember that angles on a straight line add up to 180°.

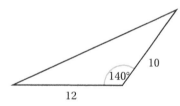

Let the resultant force be F.

Use the cosine rule.

$F^2 = 144 + 100 - 2 \times 12 \times 10 \times \cos 140°$

$\quad = 144 + 100 - 240 \cos 140°$

$\quad = 427.85$

$F = \sqrt{427.85} = 20.685 \, \text{N}$

Correct to 1 decimal place, the magnitude of the resultant force is 20.7 N.

Now that the magnitudes of all three sides are known, the sine rule can be applied to find the angle that the resultant force makes with O_x (or the cosine rule could be applied again).

In order to apply the cosine rule to find the angle, it would normally be rearranged as $\cos A = \dfrac{b^2 + c^2 - a^2}{2bc}$.

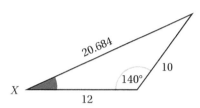

Let the angle the resultant force makes with O_x be X and apply the sine rule.

$$\frac{\sin X}{10} = \frac{\sin 140°}{20.684}$$

$$\sin X = \frac{10 \sin 140°}{20.684} = 0.311$$

$$X = \sin^{-1}(0.311) = 18.105°$$

Correct to 1 decimal place, the direction of the resultant force from O_x is 18.1°.

Exercise 1.3A

1 Forces A and B act along O_x and O_y, respectively, with magnitudes 15 N and 8 N.

   a Find the magnitude of the resultant force.

   b Find the angle that the resultant force makes with O_x.

2 A particle experiences a force of 23 N south and of 38 N west.

   a Find the magnitude of the resultant force.

   b Find the bearing of the resultant force.

3 Force G of magnitude 9 N acts along O_x. Force H of magnitude 7 N acts at 50° anticlockwise from O_x.

   a Show that the resultant force is 14.5 N.

   b Show that the resultant force acts at an angle of 21.7° to O_x.

4 Force U of magnitude 11 N acts along O_x and force V of magnitude 7 N acts at 110° anticlockwise from O_x, as shown in the diagram.

Find the magnitude and direction of the resultant force.

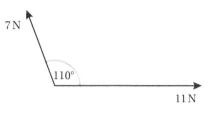

5 Forces P and Q, of magnitudes 3 N and 4 N, act along O_x and O_y, respectively. Force R, of magnitude 5 N, acts at 60° to O_x as shown in the diagram.

Hita has been asked to find the magnitude and direction of the resultant force.

Hita has written out this solution.

$a^2 = 3^2 + 5^2 - 2 \times 3 \times 5 \cos 120° = 49$

$a = 7$

$\dfrac{\sin X}{5} = \dfrac{\sin 120°}{7}$

$X = 38.2°$

$a^2 = 7^2 + 4^2 - 2 \times 7 \times 4 \cos 128.2°$

Magnitude = 9.98 N

$\dfrac{\sin Z}{4} = \dfrac{\sin 128.2°}{9.98}$

$Z = 18.4°$

Angle = 56.6°'

a Explain each part of Hita's solution.

b Write out a more efficient solution for this problem.

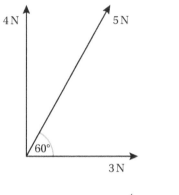

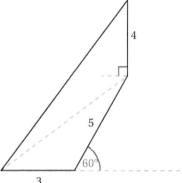

6 Force C of 18 N acts along O_x. Force D acts at 52° to the horizontal. The magnitude of the resultant force is 28 N, as shown in the diagram.

a Find the magnitude of D.

b Find the angle that the resultant force makes with O_x.

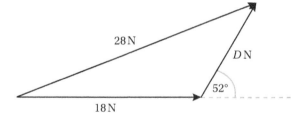

7 The force P acts along O_x. The forces Q, R and S act at 40°, 130° and 220°, respectively, to O_x, each measured anticlockwise from O_x. The forces P and Q both have magnitude 3 N and the forces R and S both have magnitude 5 N.

a Show that the resultant force acts at approximately 124° anticlockwise from O_x.

b Show that the magnitude of the resultant force is approximately 3 N.

8 A toy car experiences a force of $2x$ N on a bearing of 000° and a second force of x N at a bearing of $(180 + \alpha)°$.

a Show that the resultant force can be written as $x\sqrt{5 - 2\cos\alpha}$.

Given that the resultant force is equal to $2x$ N,

b find the size of the angle α.

9   Three horizontal forces, R, S and T, act upon a particle. Force T acts due south. Forces R and S act on bearings of $(360 - \theta)°$ and $(90 - \theta)°$, respectively. Force R is twice the magnitude of force S.

   a   Show that $\tan\theta = \frac{1}{2}$.

   b   Show that $T = S\sqrt{5}$.

10   A boat is travelling on a bearing of 050° with a driving force of 500 N.

   The wind, blowing at a force of 100 N on a bearing of 160°, is taking the boat off course.

   a   Show that the new bearing is approximately 061°.

   b   Find the direction that the boat needs to take such that, after the wind has been taken into account, the boat will travel on a bearing of 050°.

1.4 Coefficient of friction

Consider a particle sitting at rest on a horizontal surface and what happens as the surface is rotated in the vertical plane. Initially, the angle between the surface and the horizontal is zero, the particle is stationary and the forces are in equilibrium. As the angle increases, the weight of the particle has a component that should pull the particle down the slope and the only force preventing the particle from slipping is friction. As the angle increases, the frictional force required to prevent slippage increases. At a certain point, the particle will find itself on the point of slipping down the slope, and any further increase in angle will cause the particle to move down the slope.

The contact force between the particle and the surface can be represented by the frictional force and the normal reaction force. There is a relationship between the friction and reaction forces which is dependent upon the angle, as well as other factors such as the interaction between the particle and the slope (for example, a car and a hill). This relationship is given by $F \leqslant \mu R$, where μ is the **coefficient of friction**. μ generally takes a value between 0 and 1, although it is possible for μ to take a value greater than 1.

If a surface is rough, then there will be a frictional force opposing the motion. However, if the frictional force is negligible then the surface may be modelled as smooth, which means that the friction force can be assumed to be zero and the reaction force is the only component of the contact force. In reality, no surface is completely smooth.

Note that when a particle is at rest on a horizontal surface with no other external forces, the reaction force is equal and opposite to the weight, in accordance with Newton's third law. The frictional force is zero, which is of course less than the reaction force. As long as F is less than μR, there will be no motion.

When the particle is on the point of slipping or has begun to move, $F = \mu R$. This is called **limiting equilibrium**.

KEY INFORMATION

In general, $F \leqslant \mu R$, (where μ is the coefficient of friction). For all rough surfaces, $0 < F \leqslant \mu R$. For all smooth surfaces, $F = 0$.

KEY INFORMATION

$F = \mu R$ represents limiting equilibrium.

For a system in limiting equilibrium, where no forces other than reaction, friction and weight act upon a particle on a rough slope inclined at θ to the horizontal, then $\mu = \tan \theta$. This can be proved as follows.

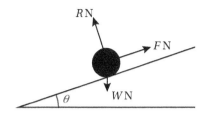

Resolve parallel to the slope.

R($\nearrow$)

$F = W \sin \theta$

Resolve perpendicular to the slope.

R($\nwarrow$)

$R = W \cos \theta$

Substitute for F and R in $F = \mu R$.

$W \sin \theta = \mu \times W \cos \theta$

Hence

$$\mu = \frac{W \sin \theta}{W \cos \theta} = \tan \theta$$

Example 8

An 18 N block is pulled by a light and inelastic taut horizontal cord on a rough horizontal table. The block is on the point of slipping. The coefficient of friction between the block and the table is 0.3. Find the tension in the cord.

Solution

The force diagram has four forces: the weight of 18 N, the reaction force, the frictional force and the tension force. If you assume that the block is being pulled to the right, then the friction acts to the left, in the opposite direction.

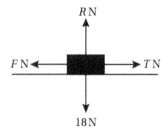

Since the particle is on the point of slipping, it is in limiting equilibrium and $F = \mu R$.

Resolve horizontally to find an expression for F.

R($\rightarrow$)

$T - F = 0$

$\quad T = F$

Resolve vertically to find an expression for R.

$R(\uparrow)$

$R - 18 = 0$

Therefore

$R = 18\,\text{N}$.

Substitute for F and R in the formula $F = \mu R$.

$T = 0.3 \times 18$

$\quad = 5.4\,\text{N}$

Example 9

A 20 N particle is held in limiting equilibrium by a horizontal force 14 N on a rough slope which is inclined at an angle of 17° to the horizontal. The particle is on the point of slipping up the slope. Calculate the coefficient of friction.

Solution

The force diagram has four forces: the weight of 20 N, the reaction force, the frictional force and the horizontal 14 N force. Since the particle is on the point of slipping up the slope, the friction acts down the slope.

> Make sure that you read the question carefully so that you draw the frictional force in the correct direction, since friction can act either up or down the slope.

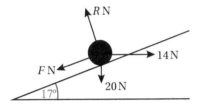

Since the particle is in limiting equilibrium, $F = \mu R$.

Resolve parallel to the slope to find an expression for F.

$R(\nearrow)$

$14 \cos 17° - F - 20 \sin 17° = 0$

Therefore

$F = 14 \cos 17° - 20 \sin 17°$.

Resolve perpendicular to the slope to find an expression for R.

$R(\nwarrow)$

$R - 20 \cos 17° - 14 \sin 17° = 0$

Therefore

$R = 20\cos 17° + 14\sin 17°$.

From $F = \mu R$, $\mu = \dfrac{F}{R}$.

$\mu = \dfrac{14\cos 17° - 20\sin 17°}{20\cos 17° + 14\sin 17°} = 0.325$

Example 10

A force of magnitude P N is applied to a particle of weight

60 N on a plane inclined at θ to the horizontal, where $\sin\theta = \dfrac{11}{16}$.

P acts parallel to the plane on a line of greatest slope up the plane. The coefficient of friction between the plane and the particle is 0.12. The particle is in limiting equilibrium.

Find the range of values for P.

Solution

The frictional force can act either up the plane, when the particle is on the point of slipping down, or up the plane, when the particle is on the point of slipping up. Draw a diagram for each.

Slipping down:

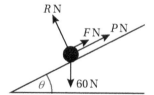

R(↗)

$P + F - 60\sin\theta = 0$

Hence $F = 60\sin\theta - P$

R(↖)

$R - 60\cos\theta = 0$

Hence $R = 60\cos\theta$

Substitute into $F = \mu R$.

$60\sin\theta - P = \mu \times 60\cos\theta$

By trigonometry, when $\sin\theta = \dfrac{11}{61}$, $\cos\theta = \dfrac{60}{61}$.

Therefore

$$60 \times \frac{11}{61} - P = 0.12 \times 60 \times \frac{60}{61}$$

$$P = 60 \times \frac{11}{61} - 0.12 \times 60 \times \frac{60}{61} = 3.74 \text{ N}.$$

Slipping up the plane:

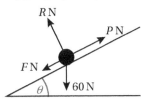

R($\nearrow$)

$P - F - 60 \sin \theta = 0$

Hence $F = P - 60 \sin \theta$.

R($\nwarrow$) $R = 60 \cos \theta$ as before.

Substitute into $F = \mu R$.

$P - 60 \sin \theta = \mu \times 60 \cos \theta$

$$P - 60 \times \frac{11}{61} = 0.12 \times 60 \times \frac{60}{61}$$

$$P = 60 \times \frac{11}{61} + 0.12 \times 60 \times \frac{60}{61} = 17.9 \text{ N}$$

Hence the range of values for P is given by $3.74 \text{ N} < P < 17.9 \text{ N}$.

Exercise 1.4A

1   A garden ornament of weight 80 N sits at rest on a concrete path inclined at 28° to the horizontal, as shown in the diagram. The ornament is in limiting equilibrium.

Find the coefficient of friction between the ornament and the path.

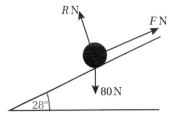

2   A car of weight 6 kN is driven up a road at a constant speed. The road is modelled as a rough surface inclined at 8° to the horizontal. Given that the coefficient of friction between the car and the road is 0.12, find the magnitude of the driving force exerted by the engine of the car.

3   A small suitcase of weight 50 N is pulled along the floor of an airport at a steady speed by mean of a light inextensible cord inclined at 20° to the horizontal, as shown in the diagram. The coefficient of friction between the suitcase and the floor is $\frac{1}{4}$.

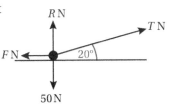

a   Show that the tension in the cord is approximately 12.2 N.

b   How would the model change if the floor was smooth?

4 A cyclist of weight 650 N is freewheeling at a steady speed down a hill inclined at an angle α to the horizontal on a bicycle of weight 280 N. Given that the coefficient of friction between the bicycle and the hill is 0.3, find the value of α, correct to 1 decimal place. The cyclist can be modelled as a particle.

5 A body of weight 32 N is held in limiting equilibrium on a rough plane inclined at 25° to the horizontal by a horizontal force P. Find the magnitude of P when the body is on the point of slipping down the plane, given that $\mu = 0.15$.

6 A piano of weight W N is being pushed up a rough slope at a constant speed. The pushing force of 560 N acts on the piano up the line of greatest slope of the plane. The coefficient of friction between the slope and the piano is $\frac{2}{7}$ and the slope is inclined at $\sin^{-1}\frac{1}{10}$ to the horizontal. Find the weight of the piano.

7 A car of weight 8500 N is on the point of slipping up a rough hill when a force P N is applied to the car along the line of greatest slope of the plane. The hill is inclined at an angle of 50° to the horizontal.

    a Find the value of μ if the force P is 15 kN.

    Given instead that $\mu \geqslant 1$,

    b find the minimum value of P.

8 The diagram shows a metal block of mass W N being pulled up a rough slope at a steady speed by a rope with tension equal to $\frac{1}{2}W$ N. The slope is inclined at angle θ to the horizontal and the rope is inclined at angle θ to the slope. The coefficient of friction between the metal block and the slope is μ.

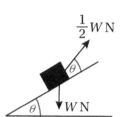

    a Show that $\mu = \dfrac{1 - 2\tan\theta}{2 - \tan\theta}$.

    b Show that $\mu > \dfrac{1}{2}$.

    c Find $\tan\theta$ when $\mu = \dfrac{3}{10}$.

9 A skateboard of weight 30 N is on a rough plane inclined at an angle of 30° to the horizontal, held at rest by a force of P N applied parallel to the line of greatest slope of the plane. The coefficient of friction between the skateboard and the plane is 0.1.

    a Find the minimum value for P.

    b Find the maximum value for P.

    c Explain why the difference between the minimum and maximum values for P is given by $3\sqrt{3}$ N.

10 A shopping trolley is pushed up a rough slope inclined at angle α to the horizontal. When the trolley is empty it has a weight of 170 N and requires a pushing force of 40 N to travel at a steady speed. The shopping trolley can be modelled as a particle.

    a Show that the coefficient of friction is given by $\dfrac{4 - 17\sin\alpha}{17\cos\alpha}$.

b Given that $\tan\alpha = \frac{13}{84}$, show that the coefficient of friction is $\frac{1}{12}$.

The same trolley, containing shopping, is now pushed up the slope. A pushing force of $100\,\text{N}$ is required.

c Find the weight of the shopping in the trolley.

11 A cart is on the point of slipping down a rough plane inclined at an angle of $30°$ to the horizontal. The coefficient of friction between the cart and the plane is $\frac{2}{9}$. The cart is held in position by a light inextensible cord inclined at $15°$ to the plane, as shown in the diagram.

The cart is modelled as a particle with a weight of $210\,\text{N}$. Find the magnitude of the tension in the cord.

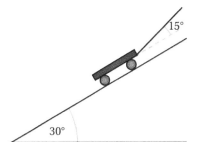

12 A rope inclined at an angle θ to the horizontal is used to pull a crate of bricks at a steady speed along a rough horizontal path. The tension in the rope is half the magnitude of the weight of the crate. The coefficient of friction between the crate and the path is μ.

a Show that μ is given by $\frac{\cos\theta}{2 - \sin\theta}$.

Given that $\cos\theta = \frac{\sqrt{5}}{3}$,

b find the value of μ.

13 A particle of weight $W\,\text{N}$ on a rough plane experiences a horizontal force of $X\,\text{N}$. The particle is on the point of slipping up the plane. The plane is inclined at an angle of α to the horizontal and the coefficient of friction is given by $\cos\alpha$.

Given, further, that $3W = X$,

a show that $10\cos^4\alpha + \cos^2\alpha = 6\cos\alpha + 1$

b verify that $30° < \alpha < 31°$.

Stop and think Compare the value of μ for a particle in limiting equilibrium on a tarmac slope at different inclinations such as $10°$, $20°$, $30°$ and $40°$. Given that for this slope $\mu \leqslant 1$, explain why this slope cannot be inclined at $50°$.

Mathematics in life and work: Group discussion

A theatre manager has asked you to design a ramp with an angle of inclination of no more than $2°$. The entrance to the theatre is $0.4\,\text{m}$ above the ground.

1 What is the minimum length the ramp could have?

2 What materials would you make the ramp from? What practical implications would you need to consider? Why are some materials more appropriate than others?

3 Why might it not be a good idea to make the ramp from a material for which μ is close to zero?

SUMMARY

> › Force is measured in newtons (N).

> › Newton's third law states that every action has an equal and opposite reaction.

> › There are five types of force that you need to be aware of:

>> › Weight acts vertically downwards.

>> › The normal reaction acts perpendicular to a surface.

>> › Tension is the force in a taut string and acts away from an object.

>> › Thrust (or compression) is the force in an inflexible rod and acts towards an object.

>> › Friction opposes motion.

> › The resultant force should be found in the direction of motion.

> › Objects are modelled as particles so their mass is assumed to be concentrated at a single point.

> › Resolving in a particular direction obtains the resultant force in that direction. It is common to resolve parallel and perpendicular to the surface. The component adjacent to the angle is given by $F\cos\theta$ and the component opposite the angle is given by $F\sin\theta$.

> › If a system is in equilibrium, then the resultant force in any direction is zero.

> › Friction acts to oppose the motion, so if a particle is on the point of slipping down a slope, then friction acts up but if the particle is on the point of moving up the slope, then the friction acts down.

> › In general, $F \leq \mu R$, (where μ is the coefficient of friction). For all rough surfaces, $0 < F \leq \mu R$. For all smooth surfaces, $F = 0$. When the particle is on the point of slipping or has begun to move, $F = \mu R$. This is called limiting equilibrium.

> › Forces can be added using Pythagoras' theorem and trigonometry or the sine and cosine rules. The positive x-axis can be written as O_x.

EXAM-STYLE QUESTIONS

1 A wardrobe is at rest on a rough slope inclined at θ to the horizontal, where $\sin\theta = \dfrac{5}{13}$. The wardrobe is on the point of slipping down the slope. Find the coefficient of friction between the wardrobe and the slope.

2 A particle is acted upon by three horizontal forces, P N, Q N and 25 N, as shown in the diagram. The Q N force acts due east and the 25 N force acts due south. The P N force acts on a bearing of 325°.

Given that the particle is in equilibrium, find:

a the value of P

b the value of Q.

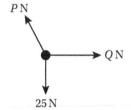

3 A peg bag of weight 1.8 N hangs in equilibrium from a washing line (made from thick rope). The rope on either side of the bag hangs at angles of 5° and 7° to the horizontal.

Find the tension in each part of the rope.

4 Angelene is pushing a small shopping trolley of weight 70 N through a supermarket at a steady speed. The floor of the supermarket is horizontal and the coefficient of friction between the trolley and the floor is $\frac{1}{12}$. The force, P N, exerted by Angelene makes an angle of 8° to the horizontal.

   a Find an expression for the reaction force in terms of P.

   b Find an expression for the frictional force in terms of P.

   c Find the value of P.

5 Three forces, F, G and H, act upon a particle in a horizontal plane. Force F has a magnitude of 24 N and acts along the positive x-axis. Force G has a magnitude of 16 N and acts at an angle of 130° to the positive x-axis as shown in the diagram. The three forces are in equilibrium.

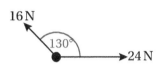

   a Find the magnitude of H.

   b Find the angle that H makes with the positive x-axis.

6 A book of weight 6 N lies on a smooth plane inclined at β to the horizontal, where $\tan \beta = \frac{3}{4}$. The book is held in equilibrium by a horizontal force of magnitude G N. Calculate the value of G.

7 Force J of magnitude 6 N acts along O_x (the positive x-axis) and force K of magnitude 9 N acts along O_y (the positive y-axis). Force L of magnitude 10 N acts at 50° to O_x, as shown in the diagram.

Find:

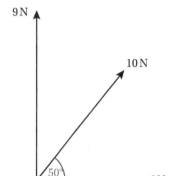

   a the magnitude of the resultant force

   b the angle the resultant force makes with O_x.

8 A particle of weight W N is held in position on a rough slope by a horizontal force of 40 N. The coefficient of friction between the particle and the slope is $\frac{1}{3}$. The slope is inclined at an angle of 35° to the horizontal. The particle is on the point of slipping up the slope. Find the value of W.

9 A block is on the point of slipping down a rough slope. The block weighs 50 N and is held in position by a horizontal force of 20 N. The slope is inclined at an angle of θ to the horizontal. The coefficient of friction between the block and the slope is 0.1.

   a Show that $\tan \theta = \frac{25}{48}$.

   b Find the magnitude of the frictional force.

10 A 24 N load is suspended from two identical cables, as shown in the diagram. The cables each make an angle of θ with the horizontal, where $\tan\theta = \frac{3}{4}$. Find the tension in each of the cables.

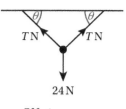

11 Force U of magnitude 5 N acts along O_x and force V of magnitude 7 N acts at an angle of 60° to O_x, as shown in the diagram. The resultant of U and V is R.

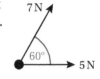

a Find the magnitude of R.

When a third force X is added, acting along O_y, the new resultant force R_2 makes an angle of 50° with O_x.

b Find the magnitude of X.

c Find the magnitude of R_2.

12 Four horizontal forces of magnitudes F N, F N, X N and 36 N act in equilibrium at a single point as shown in the diagram.

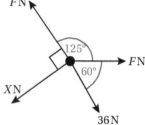

a Find the value of F.

b Find the value of X.

13 A toy helicopter is about to slip down a rough plank inclined at 20° to the horizontal. The coefficient of friction between the plank and the toy is 0.18. A horizontal force of 8 N is holding the helicopter in place.

a Find the weight of the toy helicopter.

Given that the coefficient of friction is the same when the plank is moved,

b find the minimum horizontal force required to prevent the helicopter from slipping when the plank is inclined at 30°.

14

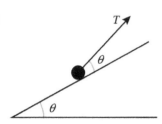

A lawnmower of weight W N sits at rest on a grass slope inclined at an angle of θ to the horizontal. The lawnmower cord is also inclined at an angle of θ to the slope. The grass slope is rough and the coefficient of friction between the lawnmower and the grass is μ. The lawnmower is on the point of slipping down the grass slope.

a Show that the tension in the cord is given by $\dfrac{W(\mu\cos\theta - \sin\theta)}{\mu\sin\theta - \cos\theta}$.

Given that the tension is 15 N, the weight is 80 N and θ is 20°,

b find the value of μ.

15 A block of weight 50 N is resting on a rough slope inclined at an angle of α to the horizontal. There is a horizontal force X N acting upon the block such that the block is on the point of sliding down the slope. The coefficient of friction between the block and the slope is μ.

Show that $X = \dfrac{50(\tan\alpha - \mu)}{1 + \mu\tan\alpha}$.

16 A particle of weight 12 N is held at rest on a rough plane by a force, P, acting at 25° to the plane. The plane is inclined at 15° to the horizontal. The coefficient of friction between the particle and the plane is 0.32. The particle is about to slip up the plane.

Find the magnitude of P.

17 A particle of weight W N is held in limiting equilibrium on a plane inclined at θ to the horizontal, where $\sin\theta = \dfrac{3}{5}$, by a force X N acting along the line of greatest slope of the plane.

The coefficient of friction between the particle and the plane is μ.

When the particle is on the point of slipping up the plane, $X = 42$ N.

When the particle is on the point of slipping down the plane, $X = 30$ N.

a Find the value of W.

b Find the value of μ.

18 A rock of weight W N is about to slip down a rough slope inclined at an angle of θ to the horizontal. The rock is held in position by a horizontal force of magnitude P N, where $5P = W$.

The coefficient of friction between the rock and the slope is $\dfrac{2}{7}$.

Show that $\tan\theta = \dfrac{17}{33}$.

19 A trolley of W N is held in limiting equilibrium on a rough plane inclined at an angle 30° to the horizontal. The trolley is on the point of rolling down the plane. The force preventing the trolley from slipping, P N, is also inclined at 30° to the horizontal, as shown in the diagram.

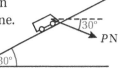

a Show that the coefficient of friction is given by $\dfrac{\sqrt{3}}{3}\left(\dfrac{W - P}{W + P}\right)$.

b Hence show that $W > P$.

20 A bead of mass 2.2 kg has been threaded onto a string of length 63 cm. The ends of the string are attached to a vertical pole at points A and B, where A is 45 cm above B. The bead is held in position by a horizontal force F N as shown in the diagram, dividing the string in the ratio 4:3 such that the part attached to A is the longer part. The tensions in each part are given by T_A N and T_B N, where $2T_A = 3T_B$.

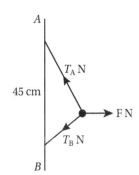

a Show that T_A is 55 N.

b Find the value of F.

21 A cupboard is being pushed up a rough ramp inclined at an angle θ to the horizontal, where $\sin \theta = 0.28$. The pushing force, P N, is directed parallel to the line of greatest slope up the ramp. The coefficient of friction between the cupboard and the ramp is given by μ.

When the cupboard is on the point of slipping down the plane, the pushing force is 5 N.

When the cupboard is on the point of slipping up the plane, the pushing force is 65 N.

a Find the weight of the cupboard.

b Hence show that $\mu = \dfrac{1}{4}$.

22 A parcel of weight W N is pulled at a steady speed across rough horizontal ground by a rope of tension T N. The coefficient of friction between the parcel and the ground is μ and the rope is inclined at angle α to the horizontal. When $\sin \alpha = \dfrac{12}{13}$, $\mu = \dfrac{5}{23}$.

a Show that $35T = 13W$.

Given that the tension is the same,

b find μ when $\sin \alpha = \dfrac{5}{13}$.

23 A table of weight W N is at rest on a rough plane. A force of P N, parallel to the line of greatest slope of the plane, is applied to the table such that the table is on the point of slipping up the slope. The value of P and the coefficient of friction, μ, between the table and the plane remain constant when the plane is angled at each of α and β to the horizontal where $\sin \alpha = \dfrac{3}{5}$ and $\sin \beta = \dfrac{8}{17}$.

a Show that $P = \dfrac{13W}{7}$.

b Hence show that $\mu = \dfrac{a}{b}$, where a and b are coprime and $a + b = 18$.

24 A block of weight W N is pulled at a steady speed across a rough horizontal plane by a cord. The cord is inclined at an angle θ to the horizontal and has a tension of T N. The coefficient of friction between the block and the plane is given by μ.

a Show that $T = \dfrac{\mu W}{\cos \theta + \mu \sin \theta}$.

When $\sin \theta = \dfrac{3}{5}$, $T = k_1 W$.

When $\sin \theta = \dfrac{4}{5}$, $T = k_2 W$.

b Given that the ratio $k_1 : k_2 = \dfrac{5}{6}$, find the value of μ.

Mathematics in life and work

A theatre has a ramp for wheelchair access. The ramp is 30 cm tall and 11.5 m long.

1 Find the angle of inclination of the ramp, correct to 1 decimal place.

The total weight of a wheelchair and its user is 1.6 kN. It is pushed up the ramp at a steady speed. The coefficient of friction between the wheelchair and the ramp is 0.14.

2 What is the minimum force that needs to be applied to the wheelchair to push it up the ramp?

For practical reasons, it is decided that this force is too high and should be reduced to 200 N. It is suggested that a different material with a lower coefficient of friction is used for the surface of the ramp.

3 Find the coefficient of friction between the wheelchair and surface of the ramp, given that the minimum force is to be 200 N.

2 KINEMATICS OF MOTION IN A STRAIGHT LINE

Mathematics in life and work

Kinematics is concerned with motion. The equations of uniformly accelerated motion describe the relationship between displacement, velocity, constant acceleration and time. Kinematics is widely applicable in a range of careers – for example:

- **›** If you were an astronaut, you would need a sound understanding of kinematics to develop the technology and techniques required for space exploration, such as, calculating the escape velocity from the Earth and timing the journey and the landing.
- **›** If you were a detective, an appreciation of kinematics might be necessary when trying to connect clues together, such as how fast a person could travel from one place to another.
- **›** If you were an athletics coach, you would need to develop a strategy for how the athlete could pace the race and when would be appropriate to accelerate, decelerate or run at a constant speed.

- **›** If you ran a taxi cab firm, it would be useful to use kinematics to ensure that you deliver the most efficient service for customers by sending the taxi that will arrive soonest.
- **›** If you were responsible for developing the timetables for an integrated transport system in a major city, you would need to be able to model the movement of buses and trains between their bus stops and railway stations.

In this chapter, you will consider kinematics applied to the design of integrated transport systems.

LEARNING OBJECTIVES

You will learn how to:

- **›** understand the concepts of distance and speed as scalar quantities, and of displacement, velocity and acceleration as vector quantities
- **›** sketch and interpret displacement–time graphs and velocity–time graphs, and to appreciate that:
  - **›** the area under a velocity–time graph represents displacement
  - **›** the gradient of a displacement–time graph represents velocity
  - **›** the gradient of a velocity–time graph represents acceleration
- **›** use differentiation and integration with respect to time to solve simple problems concerning displacement, velocity and acceleration
- **›** use appropriate formulae for motion with constant acceleration in a straight line.

LANGUAGE OF MATHEMATICS

Key words and phrases you will meet in this chapter:

acceleration, constant of integration, decelation (decelerate), differentiation, displacement, displacement–time graph, distance, gravity, integration, scalar, SI units, speed, vector, velocity, velocity–time graph

PREREQUISITE KNOWLEDGE

You should already know how to:

> draw and interpret graphs for functions of the form $ax + b$ and $x^2 + ax + b$

> find the gradient of a straight line

> carry out calculations involving the area of a rectangle, triangle and trapezium and compound shapes derived from these

> solve simple linear equations in one unknown

> solve quadratic equations by factorisation, completing the square or by use of the formula

> interpret and use graphs in practical situations, including travel graphs and conversion graphs

> draw graphs from given data

> apply the idea of rate of change to kinematics involving distance–time and speed–time graphs, acceleration and deceleration

> calculate distance travelled as area under a linear speed–time graph

> apply differentiation to gradients

> use definite integration to find the area of a region.

You should be able to complete the following questions correctly:

1 One of the equations of uniformly accelerated motion is $v^2 = u^2 + 2as$.

  a Find v when $u = 40$, $a = 18$ and $s = 49$.

  b Find s when $v = 148$, $u = 48$ and $a = 9.8$.

2 Find the area of a trapezium with parallel sides of 19 cm and 34 cm and a perpendicular height of 22 cm.

3 **a** Solve the equation $2(T + 2) = \frac{1}{2}(3T + 14)$.

  b Solve the equation $2T^2 = 5T + 12$.

4 A function is given by $f(x) = 6x^2 - 7x + 8$.

  a Differentiate $f(x)$ with respect to x.

  b Integrate $f(x)$ with respect to x.

2.1 The language of kinematics

Consider a car waiting at traffic lights. When the lights turn green, the initial speed of the car is zero. As the driver pushes down on the accelerator pedal, the car starts to move. The faster the **acceleration**, the faster the speed increases and the further the car travels in the same amount of time. Here, **distance**, **speed**, acceleration and time are all interrelated.

These quantities are connected by the equations of uniformly accelerated motion which involve **displacement** (s), initial **velocity** (u), final velocity (v), acceleration (a) and time (t). These equations are described in greater depth in **Section 2.2**. **Displacement**, **velocity** and acceleration are **vector** quantities.

Note that distance and speed are **scalar** quantities whereas displacement and velocity are vector quantities. Acceleration is a vector quantity but the same word is also used to describe the scalar quantity. In this unit, motion only ever takes place along a straight line in one dimension, so all that matters is that the sign of the displacement, velocity or acceleration (positive or negative) is correct.

SI units

SI units are an internationally recognised system of units from which all other units can be derived. The SI unit for distance is the metre and for time it is the second. The units for all the quantities in the table below can be described using just these two units: metres (m) and seconds (s).

Term	Definition	SI unit
Position	The position of an object is where it is compared with the origin, O.	The SI unit for distance travelled, displacement and position is the metre (m).
Displacement	The displacement of an object is how far the object is from its original position. For example, if an object starts at 2 metres from O and finishes at -5 metres from O, then its displacement is -7 metres. If the motion is along a horizontal line, you usually define left to right as the positive direction.	
Distance travelled	The distance travelled is how far the object has moved. For example, if an object travels 9 metres from O in the negative direction, then 5 metres back, then the displacement is -4 metres but the distance travelled is 14 metres.	
Velocity	The velocity is how quickly an object is travelling in a certain direction. For example, if an object travels a displacement of -10 metres in 5 seconds, then its velocity is -2 metres per second (m s^{-1}). If an object changes direction from positive to negative (or negative to positive) there will be an instant when the velocity is zero. This is called instantaneous rest.	The SI unit for both speed and velocity is the metre per second (m s^{-1}) since they are the rate of change of distance and displacement, respectively. The unit m s^{-1} means the same as m/s (metres per second). The index of $^{-1}$ shows that you are dividing by time.
Speed	Speed is how quickly an object is travelling but the direction does not matter. Hence the speed in the previous example is 2 metres per second (m s^{-1}) even though the particle is travelling in a negative direction. Note that two particles with velocities of $5\,\text{m s}^{-1}$ and $-5\,\text{m s}^{-1}$ have the same speed but are travelling in opposite directions.	

Term	Definition	SI unit
Acceleration	Acceleration describes how quickly the speed (or velocity) is changing. For example, if an object increases in speed from $4\,\text{m s}^{-1}$ to $10\,\text{m s}^{-1}$ in 2 seconds, then its acceleration is 3 metres per second per second (m s^{-2}). Note that a particle with a positive velocity but a negative acceleration will come to instantaneous rest before travelling in the opposite direction. When the velocity and acceleration are different signs, the object will be **decelerating**.	The SI unit for acceleration is the metre per second per second (m s^{-2}) since it is the rate of change of speed or velocity. The unit m s^{-2} means the same as m/s^2 (metres per second squared). This represents the change of velocity in m s^{-1} each second.

Example 1

A girl walks along a straight line. She walks 35 m in the positive direction, then turns round and walks 56 m in the negative direction.

a Find the total distance travelled by the girl.

b Find the girl's displacement at the end of her walk from her starting point.

Solution

Draw a diagram.

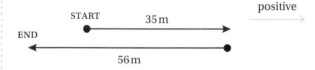

a The girl has walked 35 m in one direction and then 56 m in another.
The total distance is given by 35 + 56 = 91 m.

b The girl has walked 35 m in the positive direction and then 56 m in the negative direction.
The overall displacement is given by 35 – 56 = –21 m.

> **Stop and think**
> Why does the direction not matter when working out the total distance?
> If you travel further in the negative direction than the positive direction, why will your overall displacement be negative?

Exercise 2.1A

1 A particle travels 30 m in the negative direction, then 20 m in the positive direction.
Find the overall distance travelled and the displacement from its original position.

2 A rocket is launched vertically upwards from the ground. It reaches a maximum height of 73 m, then returns to the ground.

    a Find the total distance travelled by the rocket.

    b Find the overall displacement of the rocket.

3 A ball is launched upwards from the top of a 15 m tall building to a maximum height of 40 m, before falling to the ground. Taking upwards as the positive direction, find the overall distance travelled by the ball and the displacement of the ball from its original position.

4 A snail crawls around the perimeter of a rectangle with sides of 7 m and 4 m.

    a Find the total distance travelled by the snail.

    b Find the overall displacement of the snail at the end of its journey from its starting point.

5 A cyclist travels from her house to her grandparents' house. She cycles 16 km east, then 16 km south.

    a Find the bearing of her grandparents' house from her own house.

    b Find her overall displacement.

    c How much further than her displacement did she cycle?

6 A boy walks 32 m west, then 24 m south.

    a Find the total distance travelled by the boy.

    b Find the displacement of the boy from his starting point.

    c If it took the boy 28 seconds, what was his speed?

7 A yacht sails at a constant speed for 20 minutes on a bearing of 120°, then for another 20 minutes on a bearing of 240°. After the journey, the yacht is 10 km from its starting point.

    a Find the distance sailed by the yacht.

    b Find the speed at which the yacht was sailing.

8 An ant walks along the lines of a coordinate grid from the point (38, –4) to the point (–2, 5). The ant takes the quickest journey possible along the lines.

    a Find the total distance travelled by the ant.

    b Find the displacement of the ant at the end of its journey from the start of its journey, writing the direction as a bearing to the nearest integer.

    c If the ant walks one square per second, how much longer does it take the ant to walk along the lines than taking the direct route?

9 A bus drives 15 km on a bearing of 060°, then 36 km on a bearing of 150°.

    a Find the difference between the length of the journey taken by the bus and the direct route.

    b Find the overall displacement of the bus, writing the direction as a bearing to the nearest integer.

2.2 Equations of uniformly accelerated motion

Consider a car travelling at a constant acceleration. It starts with an initial velocity of $u\,\mathrm{m\,s^{-1}}$ and after $t\,\mathrm{s}$ it has a final velocity of $v\,\mathrm{m\,s^{-1}}$. In the velocity–time graph below you can see that this is represented as a straight line with a constant gradient. u is the initial velocity, v is the final velocity and t is the time taken. Its constant acceleration is given by $a\,\mathrm{m\,s^{-2}}$ and it has travelled $s\,\mathrm{m}$.

Note that of the five variables (s, u, v, a and t), time is the only variable that is *scalar*, not a vector. Whereas the other quantities can take negative values and motion can take place forwards or backwards, time can only take positive values.

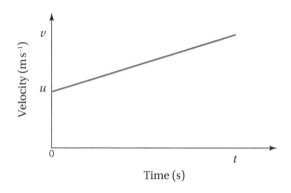

The five quantities (displacement (s), the initial velocity (u), the final velocity (v), the constant acceleration (a) and the time (t)) are related by the equations of uniformly accelerated motion. Each equation can be derived from the definition of acceleration (the rate of change of velocity).

Acceleration is the rate of change of velocity; the gradient on the graph.

From the graph, the acceleration is given by

$$a = \frac{v - u}{t}.$$

This can be rearranged to make v the subject.

$$at = v - u$$

$$v = u + at$$

An alternative method for deriving the equations of uniformly accelerated motion is to integrate constant acceleration twice with respect to time.

Another equation can be derived by considering the average speed.

The average speed can be written in terms of the displacement (s) and the time (t).

Average speed $= \dfrac{s}{t}$.

The average speed can also be written in terms of the initial velocity (u) and the final velocity (v).

Average speed $= \dfrac{1}{2}(u + v)$.

Put these expressions equal to each other.

$$\frac{s}{t} = \frac{1}{2}(u + v)$$

Multiply by t.

$s = \frac{1}{2}(u + v)t$

By substituting for $v = u + at$ into $s = \frac{1}{2}(u + v)t$, you can derive a third equation, $s = ut + \frac{1}{2}at^2$.

$s = \frac{1}{2}(u + v)t$

Substitute for v.

$s = \frac{1}{2}(u + u + at)t$

Simplify.

$s = \frac{1}{2}(2u + at)t$

Expand.

$s = \frac{1}{2} \times 2ut + \frac{1}{2} \times at^2$

Simplify.

$s = ut + \frac{1}{2}at^2$

By making t the subject of $s = \frac{1}{2}(u + v)t$ and substituting for t in the equation $v = u + at$, you can derive the fourth equation: $v^2 = u^2 + 2as$.

$s = \frac{1}{2}(u + v)t$

Multiply by 2.

$2s = (u + v)t$

Divide by $(u + v)$.

$t = \frac{2s}{u + v}$

$v = u + at$

Substitute for t.

$v = u + \frac{2as}{u + v}$

Subtract u.

$v - u = \frac{2as}{u + v}$

Multiply by $(u + v)$.

$(v - u)(u + v) = 2as$

Expand.

$v^2 - u^2 = 2as$

Add u^2.

$v^2 = u^2 + 2as$

In summary, the equations are

$v = u + at$ (no s)

$v^2 = u^2 + 2as$ (no t)

$s = \frac{1}{2}(u + v)t$ (no a)

$s = ut + \frac{1}{2}at^2$ (no v)

$s = vt - \frac{1}{2}at^2$.

These five equations are often called the SUVAT equations.

Each of the equations has four of the variables s, u, v, a and/or t.

KEY INFORMATION

You should learn these equations and make sure that you can rearrange them.

Stop and think How can you use SI units to show that the terms v and at have the same dimensions and that the terms v^2 and as have the same dimensions?

Example 2

A ball travels 143 m in 13 s at a constant acceleration. Given that its final velocity is $14\,\text{m s}^{-1}$, find its initial velocity.

Solution

You are told that:

$s = 143\,\text{m}$

$t = 13\,\text{s}$

$v = 14\,\text{m s}^{-1}$

u is unknown.

Write down the equation you are going to use.

$s = \frac{1}{2}(u + v)t$

Substitute for s, t and v.

$143 = \frac{1}{2}(u + 14) \times 13$

Rearrange the equation to make u the subject.

Divide both sides by 13.

$11 = \frac{1}{2}(u + 14)$

Multiply both sides by 2.

$22 = u + 14$

Subtract 14.

$u = 8\,\text{m s}^{-1}$

What information have you been given? Write it out using s, u, v, a and t.

What is the only variable you do not need for this question? How does that help you choose which equation to use?

Remember to include units with your answer.

Example 3

A train travelling at an initial velocity of $130\,\text{m}\,\text{s}^{-1}$ decelerates for half a minute with a constant deceleration. It travels 2.1 km during this time.

a Find the magnitude of the deceleration.

b Find the final velocity of the train.

Solution

a You are told that:

$u = 130\,\text{m}\,\text{s}^{-1}$

$t = 30\,\text{s}$

$s = 2100\,\text{m}$

> Ensure that you convert measurements into SI units before you start.

a is unknown.

Write down the equation.

$s = ut + \dfrac{1}{2}at^2$

> Choose the equation which has the letters u, t, s and a.

Substitute.

$2100 = 130 \times 30 + \dfrac{1}{2} \times a \times 30^2$

Simplify the terms.

$2100 = 3900 + 450a$

Make a the subject.

$2100 - 3900 = 450a$

$-1800 = 450a$

$a = -4\,\text{m}\,\text{s}^{-2}$

> How can you check that your answer makes sense? Would you expect a negative answer for the acceleration? How do you know that your answer should be negative?

b Since you know the values of u, t, s and a, you can now use any of the equations that contain v to find v. The simplest is $v = u + at$.

$v = u + at$

$= 130 - 4 \times 30$

$= 130 - 120$

$= 10\,\text{m}\,\text{s}^{-1}$

Example 4

A car travels along a straight line ABC at a constant acceleration. $AB = 27\,\text{m}$ and $BC = 85\,\text{m}$. The car takes 3 seconds to get from A to B and another 5 seconds to get from B to C.

a Find the acceleration of the car.

b Find the speed of the car at A.

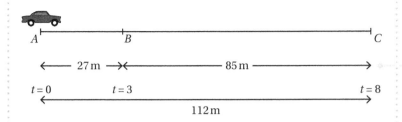

Drawing a diagram is often the best way to understand a situation.

Solution

a Write out what you know for *AB* and *AC*.

For *AB*: $s = 27$ m and $t = 3$ s

For *AC*: $s = 27 + 85 = 112$ m and $t = 3 + 5 = 8$ s

Use the equation that has s, t, u and a.

For *AB*:

$$s = ut + \frac{1}{2}at^2$$

Substitute $s = 27$ and $t = 3$ into the equation.

$$27 = u \times 3 + \frac{1}{2} \times a \times 3^2$$

Simplify the equation.

$$27 = 3u + \frac{9}{2}a$$

Multiply by 2 so that a does not have a fractional coefficient.

$$54 = 6u + 9a$$

Divide through by common factor 3. Label this equation ①.

$$18 = 2u + 3a \qquad\qquad ①$$

For *AC*:

$$s = ut + \frac{1}{2}at^2$$

Substitute $s = 112$ and $t = 8$ into the equation.

$$112 = u \times 8 + \frac{1}{2} \times a \times 8^2$$

Simplify.

$$112 = 8u + 32a$$

Divide though by common factor 8. Label this equation ②.

$$14 = u + 4a \qquad\qquad ②$$

Double equation ② so that the equations ① and ② have the same coefficient for u. Label this equation ③.

$$28 = 2u + 8a \qquad\qquad ③$$

$$18 = 2u + 3a \qquad\qquad ①$$

Subtract equation ① from equation ③:

$$10 = 5a$$

$$a = 2\,\text{m s}^{-2}$$

Why is it best to consider the journeys from *A* to *B* and from *A* to *C* rather than from *B* to *C*?

Since you only have two pieces of information (displacement and time), you need to use simultaneous equations.

b Substitute $a = 2$ into equation ① (in fact, any equation will do).

$18 = 2u + 3 \times 2$

Solve the equation.

$18 = 2u + 6$

$12 = 2u$

$u = 6\,\text{m s}^{-1}$

Stop and think What type of equation would you have to solve if you were given the values of s, u and a and needed to find the value of t? Is there a way to avoid having to solve this type of equation? What might you need to remember if you were going to use a different method?

Exercise 2.2A

1 **a** Find s, given that $u = 70\,\text{m s}^{-1}$, $a = -3\,\text{m s}^{-2}$ and $t = 10\,\text{s}$.

b Find s, given that $u = 15\,\text{m s}^{-1}$, $v = 29\,\text{m s}^{-1}$ and $t = 9\,\text{s}$.

c Find a, given that $u = 3\,\text{m s}^{-1}$, $v = 38\,\text{m s}^{-1}$ and $t = 7\,\text{s}$.

d Find s, given that $u = 22\,\text{m s}^{-1}$, $a = 6\,\text{m s}^{-2}$ and $v = 28\,\text{m s}^{-1}$.

e Find t, given that $v = -28\,\text{m s}^{-1}$, $a = -7\,\text{m s}^{-2}$ and $s = 0\,\text{m}$.

2 A particle has an initial velocity of $24\,\text{m s}^{-1}$ and a constant acceleration of $5\,\text{m s}^{-2}$.

a Find its velocity after it has travelled $10\,\text{m}$.

b Find the time taken to travel $10\,\text{m}$.

3 A particle has an initial velocity of $60\,\text{m s}^{-1}$ and constant deceleration of $8\,\text{m s}^{-2}$.

a Find the velocity of the particle after 10 seconds.

b Find the time at which the particle is at instantaneous rest.

4 From a standing start, a cyclist wins a sprint race in 40 seconds. She has a constant acceleration of $0.5\,\text{m s}^{-2}$.

a How fast was she travelling when she crossed the finish line?

b What distance was the race?

5 A bullet is fired from a gun at $50\,\text{m s}^{-1}$ and travels $1500\,\text{m}$ before it stops.

a Find the deceleration of the bullet and the length of time it is in flight.

b What assumptions must you make to answer this question?

6 **a** A particle at rest suddenly accelerates at $4\,\mathrm{m\,s^{-2}}$. How long does it take the particle to travel $128\,\mathrm{m}$?

   b A particle travelling at $180\,\mathrm{m\,s^{-1}}$ starts to decelerate at a constant $3\,\mathrm{m\,s^{-2}}$. How many minutes does it take to return to its original position?

   c A particle with an initial velocity of $5\,\mathrm{m\,s^{-1}}$ and a constant acceleration of $3\,\mathrm{m\,s^{-2}}$ takes T seconds to travel $84\,\mathrm{m}$. Show that $3T^2 + 10T - 168 = 0$ and hence find the value of T.

7 A coin is rolling down a hill with a constant acceleration of $0.25\,\mathrm{m\,s^{-2}}$. It travels $930\,\mathrm{m}$ in a minute. What was the initial speed of the coin and what was its speed after one minute?

8 A car accelerates uniformly from rest to $60\,\mathrm{km\,h^{-1}}$ in eight seconds, then maintains a steady speed.

   a How long does it take the car to travel $200\,\mathrm{m}$?

   b How long does it take the car to travel one kilometre?

9

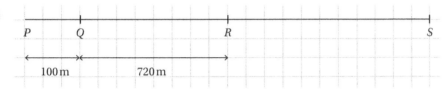

A particle moves at a constant acceleration along a straight line $PQRS$, passing Q five seconds after P and passing R fifteen seconds after Q. PQ is $100\,\mathrm{m}$ and QR is $720\,\mathrm{m}$.

   a Find the acceleration of the particle.

   b Find the velocity of the particle at Q.

   c Given that RS is $830\,\mathrm{m}$, how long does it take the particle to travel from Q to S?

10 After a windy night, a road has a pile of debris blocking the way. A motorist driving towards the debris notices it when he is $80\,\mathrm{m}$ away, travelling at $20\,\mathrm{m\,s^{-1}}$ and with a constant acceleration of $2\,\mathrm{m\,s^{-2}}$. As soon as he engages the brakes, he will decelerate at a constant $4\,\mathrm{m\,s^{-2}}$.

   a If the motorist reacts immediately, how far from the debris will he stop?

   b If he takes two seconds to react, at what speed will he be driving when he crashes into the debris?

11 Particles A and B are at rest at either end of a $100\,\mathrm{m}$ road. They set off instantaneously towards each other with constant accelerations of $6\,\mathrm{m\,s^{-2}}$ and $2\,\mathrm{m\,s^{-2}}$, respectively.

   a Where will they meet?

   b When will they meet?

   c At what speed will each be travelling when they meet?

12 A ball is launched along a straight line with an initial velocity at $13\,\mathrm{m\,s^{-1}}$ at a constant acceleration of $-2\,\mathrm{m\,s^{-2}}$. At what times is the ball $30\,\mathrm{m}$ from its starting point?

13 J passes O at $2\,\mathrm{m\,s^{-1}}$ and a constant acceleration of $1\,\mathrm{m\,s^{-2}}$. When J is $30\,\mathrm{m}$ from O, K sets off from rest from O at a constant acceleration of $0.8\,\mathrm{m\,s^{-2}}$. When K reaches a velocity of $4\,\mathrm{m\,s^{-1}}$, L passes O at $3\,\mathrm{m\,s^{-1}}$ and a constant acceleration of $0.5\,\mathrm{m\,s^{-2}}$. If J, K and L are all travelling in the same direction along the same straight line, find the distance between J and K when L has a velocity of $7\,\mathrm{m\,s^{-1}}$.

14 A body moves along the straight line *EFGH* with a constant acceleration. It takes the body seven seconds to travel between *E* and *F*, three seconds to travel the 66 m between *F* and *G* and four seconds to travel between *G* and *H*, where it comes to rest.

   a Find the deceleration of the body.

   b Find the velocity of the body at *E*.

PS 15 A cyclist passes a garage, a restaurant and a library, in that order, along a straight road, maintaining a constant acceleration of $0.15\,\mathrm{m\,s^{-2}}$ throughout and passing the library at $9\,\mathrm{m\,s^{-1}}$. The library is 240 m from the restaurant. The cyclist takes twice as long to cycle from the restaurant to the library as she does to cycle from the garage to the restaurant. How far is the library from the garage?

Mathematics in life and work: Group discussion

You work for a bus company and have responsibility for ensuring that the timetables are accurate. One part of one of your company's routes passes through a busy high street. Two bus stops, labelled *A* and *B* on your map, are 150 m apart. There is one set of traffic lights, 50 m after *A*, but other than that the traffic flows smoothly. There is a bus is scheduled to leave *A* at 10.15 am and you need to determine the time it should arrive at *B* for the timetable.

1 What equations could you use to estimate the length of time it would take to travel from bus stop *A* to the traffic lights or from the traffic lights to bus stop *B*? What limitations do these equations have in this situation?

2 What other factors might you need to consider when working out the total journey time from one bus stop to another? For example, would it matter what time of day it was or which day of the week and why?

3 If the maximum acceleration and deceleration of the bus is $0.4\,\mathrm{m\,s^{-2}}$, what impact would this have on your decisions?

4 As a group, make a reasonable estimate for the time that should be written on the timetable for the departure of the bus from bus stop *B*.

2.3 Vertical motion

One specific example of a constant acceleration is the acceleration due to **gravity**, which pulls objects towards the centre of the Earth. At the surface of the Earth this is approximately $10\,\mathrm{m\,s^{-2}}$. All objects are pulled towards the Earth at the same acceleration, regardless of their mass.

> Although the actual value of the acceleration due to gravity varies with the distance from the centre of the Earth, you can assume it is constant over small distances.

Because the force of gravitational attraction (gravity) always acts towards the centre of the Earth, you will always represent it in a diagram as acting vertically downwards. Because displacement, velocity and acceleration are vector quantities, you will need to choose which direction to take as the positive direction. You usually define the direction in which the object starts moving as positive, so if an object is launched vertically upwards then that will be the positive direction (and $a = -10\,\mathrm{m\,s^{-2}}$), whereas if the object is dropped or thrown vertically downwards then downwards will be the positive direction (and $a = 10\,\mathrm{m\,s^{-2}}$).

Example 5

A ball is thrown vertically upwards at $30\,\mathrm{m\,s^{-1}}$ from $6\,\mathrm{m}$ above the ground.

a Find the maximum height reached by the ball.

b Find the time taken for the ball to reach its maximum height.

c Find the time taken for the ball to hit the ground.

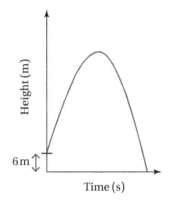

Solution

a You wish to find the maximum height, which is the distance from the ground, so you need to find the value of s. You are given u, and you know that $v = 0\,\mathrm{m\,s^{-1}}$ and $a = -10\,\mathrm{m\,s^{-2}}$, so choose the equation that has s, u, v and a. That equation is $v^2 = u^2 + 2as$.

$v^2 = u^2 + 2as$

Substitute the values into the equation.

$0 = 30^2 + 2 \times -10 \times s$

Simplify and solve for s.

$0 = 900 - 20s$

$20s = 900$

$s = 45\,\mathrm{m}$

Since the ball was thrown from $6\,\mathrm{m}$ above the ground, you need to add that to your answer, so the maximum height is $51\,\mathrm{m}$.

b For the time taken, the simplest equation to use is $v = u + at$.

Substitute and solve for t.
$$0 = 30 - 10t$$
$$10t = 30$$
$$t = 3\,\text{s}$$

c   For the time taken for the ball to hit the ground, the displacement is $-6\,$m. You know the values of s, u and a, so use $s = ut + \dfrac{1}{2}at^2$.

It is important to make sure you are using the correct sign (negative or positive) for each value. Check the direction of each vector carefully.

Write down the equation.
$$s = ut + \frac{1}{2}at^2$$

Substitute.
$$-6 = 30t + \frac{1}{2} \times -10 \times t^2$$

Simplify.
$$-6 = 30t - 5t^2$$

Because this is a quadratic equation, rearrange the terms so that they are in the form $at^2 + bt + c = 0$.
$$5t^2 - 30t - 6 = 0$$

Solve the equation by using the quadratic formula.

$$t = \frac{30 \pm \sqrt{(-30)^2 - 4 \times 5 \times (-6)}}{2 \times 5}$$

$$= 6.19\,\text{s or } -0.194\,\text{s}$$

Alternative method: Use $v^2 = u^2 + 2as$ to find v, then use $v = u + at$ to find t. Remember that v will be in the opposite direction to the direction the ball was thrown.

Because the time must be positive, it takes the ball $6.19\,$s to hit the ground.

Example 6

A ball is dropped from the top of a building.

It takes the ball $2.2\,$s to reach the ground.

Since the ball is dropped, what do you know about the initial velocity?

a   How tall is the building?

Janet's office is two-thirds of the way up the building.

b   With what velocity did the ball pass Janet's office?

Solution

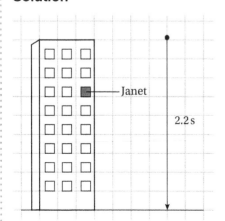

a You are told that $t = 2.2$ s. To find the height of the building you need to find the value of s. The equation that has u, a, t and s is $s = ut + \frac{1}{2}at^2$.

$$s = ut + \frac{1}{2}at^2$$

Substitute.

$$s = 0 \times 2.2 + \frac{1}{2} \times 10 \times 2.2^2$$

$$= 24.2 \text{ m}$$

> Start by choosing which direction is positive. Since the motion starts downwards, let that be the positive direction. Hence acceleration (gravity) is acting in the positive direction, and for all parts of this question, $u = 0 \text{ m s}^{-1}$ and $a = 10 \text{ m s}^{-2}$.

b Janet's office is two-thirds of the way up the building, so the ball will have travelled one-third of its total distance.

$$s = \frac{1}{3} \times 24 = 8.07 \text{ m}$$

The equation you need is the one with s, u, a and v, which is $v^2 = u^2 + 2as$.

that

Substitute into the equation.

$$v^2 = 0^2 + 2 \times 10 \times 8.07$$

$$= 161.4$$

$$v = \sqrt{161.4}$$

$$= 12.7 \text{ m s}^{-1}$$

Stop and think If a ball is dropped, how will this simplify the equation $s = ut + \frac{1}{2}at^2$? What will be the relationship between displacement and time when the ball is dropped? If it takes twice as long to hit the ground, how much taller will the building be?

Example 7

A ball is dropped from the top of a tower of height H m. If the ball had been launched vertically upwards at 16 m s^{-1}, it would have taken two seconds longer to reach the ground. Find the value of H.

Solution

Let upwards be the positive direction and let the time taken for the dropped ball to reach the ground be T.

For the dropped ball, $u = 0 \text{ m s}^{-1}$, $s = -H$m, $a = -10 \text{ m s}^{-2}$, $t = T$s.

Use $s = ut + \frac{1}{2}at^2$.

Substitute.

$$-H = 0 \times T + \frac{1}{2} \times -10 \times T^2$$

> **KEY INFORMATION**
>
> If two objects are launched in different ways, look for what their journeys have in common.

> The ball will move in different directions depending upon the situation, so there is no clear choice as to which direction to take as positive.

Simplify.

$-H = -5T^2$ ①

For the launched ball, $u = 16\,\text{m}\,\text{s}^{-1}$, $s = -H\,\text{m}$, $a = -10\,\text{m}\,\text{s}^{-2}$, $t = (T + 2)\,\text{s}$.

Use $s = ut + \frac{1}{2}at^2$.

Substitute.

$-H = 16 \times (T + 2) + \frac{1}{2} \times -10 \times (T + 2)^2$

Simplify.

$-H = 16(T + 2) - 5(T^2 + 4T + 4)$

$\quad = 16T + 32 - 5T^2 - 20T - 20$

$-H = -5T^2 - 4T + 12$ ②

Put ① = ②.

$-5T^2 = -5T^2 - 4T + 12$

Subtract $-5T^2$ from both sides.

$0 = -4T + 12$

Solve the equation.

$4T = 12$

$\ T = 3\,\text{s}$

Since $-H = -5T^2$, you can find H by substituting $T = 3$.

$-H = -5T^2$

Divide both sides by -1.

$H = 5T^2$

$\ = 5 \times 3^2$

$\ = 45$

Hence, the tower is 45 m tall.

Example 8

Particle A is launched vertically upwards from the ground at $43\,\text{ms}^{-1}$. 3 s later a second particle, B, is launched vertically upwards from the same place at $34\,\text{ms}^{-1}$.

a Find the maximum height attained by A.

b Find the distance below this maximum height at which A and B collide.

Solution

a For A, $u = 43$ ms^{-1} and $a = -10$ ms^{-2}.

At the maximum height, $v = 0$ ms^{-1}.

Using $v^2 = u^2 + 2as$

$s = \dfrac{v^2 - u^2}{2a} = \dfrac{0 - 43^2}{2 \times -10} = 92.45$ m

Maximum height attained by A is 92.5 m correct to 3 significant figures.

b Let the time that A has been in motion be T s and the time that B has been in motion be $(T - 3)$ s.

Also for B, $u = 34$ ms^{-1} and $a = -10$ ms^{-2}.

Substitute information into $s = ut + \dfrac{1}{2}at^2$ for each particle.

For A,

$s = 43T + \dfrac{1}{2} \times -10 \times T^2 = 43T - 5T^2$

For B,

$s = 34(T - 3) + \dfrac{1}{2} \times -10 \times (T - 3)^2 = 34T - 5(T - 3)^2$

When A and B collide, the displacements will be the same.

$43T - 5T^2 = 34(T - 3) - 5(T - 3)^2$

$43T - 5T^2 = 34(T - 3) - 5(T^2 - 6T + 9)$

$43T - 5T^2 = 34T - 102 - 5T^2 + 30T - 45$

$43T = 34T - 102 + 30T - 45$

$147 = 21T$

$T = 7$

When $T = 7$, $s = 43(7) - 5(7)^2 = 301 - 245 = 56$ m.

Distance = $92.45 - 56 = 36.45$.

Distance below maximum height is 36.5 m correct to 3 significant figures.

Exercise 2.3A

1 A pebble is dropped down a well. It takes 1.8 seconds until the pebble hits the water. Find the distance travelled by the pebble before it hits the water.

2 A plate falls from a table. The table is 1.2 m tall. Find the time taken for the plate to hit the ground.

3 A ball is thrown vertically downwards at 7 m s^{-1} and strikes the ground at 15 m s^{-1}. From how far above the ground was the ball thrown?

4 A stone is dropped from the summit of a 48 m cliff.

Find, correct to 3 s.f.:

a the speed of the stone when it hits the ground

b the time taken until the stone hits the ground.

5 A rocket is launched vertically upwards from a cliff at a speed of $40\,\mathrm{m\,s^{-1}}$.

It hits the ground 10 seconds later.

a Find the height of the cliff.

b Find the maximum height reached by the rocket.

c Find the time taken to reach the maximum height.

d Find the speed at which the rocket hits the ground.

e Find the time taken for the rocket to return to the same height from which it was launched.

6 A ball of weight 8 N is launched vertically upwards into the air from the ground at $25\,\mathrm{m\,s^{-1}}$.

a Find the time taken for the ball to return to the ground.

b Find the maximum height that the ball reaches above the ground.

c Find the time taken for the ball to first reach three-quarters of its maximum height.

d Is the maximum height an underestimate or overestimate? Explain your answer.

e How would the answers change if the ball had a weight of 16 N instead?

7 **a** An object is dropped from above the ground. If h is the initial height and t is the time until the object hits the ground, prove that $t = \sqrt{\dfrac{2h}{g}}$.

b If the same object had been dropped on another planet from a point 12 times higher and the gravitational acceleration had been 3 times greater, how many times longer would it have taken the object to hit the ground?

8 A stone is dropped from the top of a castle. One second later another stone is thrown downwards from the same place at $14\,\mathrm{m\,s^{-1}}$. Given that both stones hit the moat at the same time, find the height of the castle.

9 Two students, Sawda and Katie, were given this question.

'A ball is launched vertically upwards at $15\,\mathrm{m\,s^{-1}}$ from 20 m above the ground. Find the velocity with which the ball hits the ground.'

Sawda's solution:

$s = ut + \dfrac{1}{2}at^2$

$-20 = 15t - 5t^2$

$5t^2 - 15t - 20 = 0$

$t^2 - 3t - 4 = 0$

$(t - 4)(t + 1) = 0$

$t = 4 \text{ or } -1$

Since t is positive, $t = 4$ seconds

$v = u + at = 15 - 10 \times 4 = -25\,\text{m}\,\text{s}^{-1}$

Katie's solution:

$v^2 = u^2 + 2as$

$\quad = 15^2 + 2 \times (-10) \times (-20)$

$\quad = 625$

$v = 25\,\text{m}\,\text{s}^{-1}$

Whose solution is correct? Evaluate their approaches/solutions.

10 An object is launched vertically upwards at $70\,\text{m}\,\text{s}^{-1}$ from the ground. T seconds after launching, the object is higher than $165\,\text{m}$ above the ground.

    a Show that $T^2 - 14T + 33 < 0$.

    b Hence find the length of time for which the object is higher than $165\,\text{m}$ above the ground.

11 A ball is projected vertically upwards from a balcony $11.4\,\text{m}$ high. It takes the ball T seconds to reach its maximum height and 4.3 seconds for the ball to land on the ground. Find the value of T.

12 A golf ball is thrown vertically upwards from the ground with a velocity of $43\,\text{m}\,\text{s}^{-1}$.

    For how long is the golf ball more than $90\,\text{m}$ above the ground?

13 A weight is pulled vertically upwards from rest with acceleration of $0.8\,\text{m}\,\text{s}^{-2}$ by a rope from the top of a table. After four seconds, the rope snaps, at which point the weight travels only under the influence of gravity. How long after the rope snaps does the weight return to the table?

14 A stone is dropped from the top of a $37\,\text{m}$ tall building. When it is halfway down the building, another stone is thrown vertically downwards. At what speed must the second stone be thrown for both stones to hit the ground at the same time?

15 Holly can only catch a ball travelling at $5\,\text{m}\,\text{s}^{-1}$ or slower. She is standing at the top of a building, with her hands $45\,\text{m}$ above the ground. Ruby launches a ball vertically upwards from the ground at $U\,\text{m}\,\text{s}^{-1}$.

    a What are the minimum and maximum values of U if Holly is to catch the ball on her first attempt?

    b If Holly misses the ball on her first attempt, what is the maximum time before she gets a second chance?

16 A ball is dropped from the top of a tower of height $H\,\text{m}$. If the ball had been launched vertically upwards at $U\,\text{m}\,\text{s}^{-1}$, it would have taken X seconds longer to reach the ground. Show that H is given by $H = 5X^2\left(\dfrac{U - 5X}{10X - U}\right)^2$.

2.4 Displacement–time and velocity–time graphs

You saw in **Section 2.1** that velocity is the rate of change of displacement and acceleration is the rate of change of velocity. In Pure Mathematics 1 you saw how differentiation and integration could be used to find the gradient and area under a curve.

Differentiating any function gives a rate of change (whether it is with respect to distance x or time t, or any other variable). Hence there is a connection between **displacement–time graphs** and **velocity–time graphs** and the results found by calculus.

For a displacement–time graph, the gradient is the velocity (because velocity is the rate of change of displacement with respect to time).

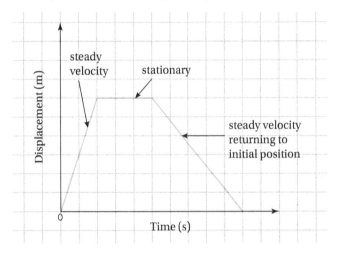

> **KEY INFORMATION**
>
> For a displacement–time graph, the gradient is the velocity.

For a velocity–time graph, the gradient is the acceleration (because acceleration is the rate of change of velocity with respect to time) and the area under the graph is the displacement (because by the fundamental theorem of calculus, differentiation and integration are inverse processes).

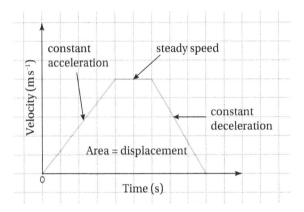

In this section, you will apply these concepts to graphs constructed from straight lines, where the constant acceleration equations apply. Then, in **Section 2.5**, you will use calculus to solve problems where displacement, velocity and acceleration are functions of time.

Example 9

A boy is on his way to school. He leaves home at 7:40 a.m. and walks $\frac{1}{2}$ km at $5\,\text{km}\,\text{h}^{-1}$ to the bus stop. The bus arrives 9 minutes after he does and travels the 12 km journey to school at an average speed of $20\,\text{km}\,\text{h}^{-1}$. School starts at 8:30 a.m.

a Draw a displacement–time graph for this journey.

b Does the boy arrive on time?

Solution

a The boy walks at $5\,\text{km}\,\text{h}^{-1}$. 5 kilometres per hour is the same as 1 kilometre every 12 minutes, so it will take the boy 6 minutes to walk $\frac{1}{2}$ km.

The boy leaves home at 7:40 a.m. and arrives at the bus stop 6 minutes later at 7:46 a.m.
The bus arrives 9 minutes later at 7:55 a.m.

The bus travels at $20\,\text{km}\,\text{h}^{-1}$. 20 kilometres per hour is the same as 1 kilometre every 3 minutes, so it will take the bus 36 minutes to travel 12 kilometres.

The bus arrives 36 minutes after 7:55 a.m. at 8:31 a.m.

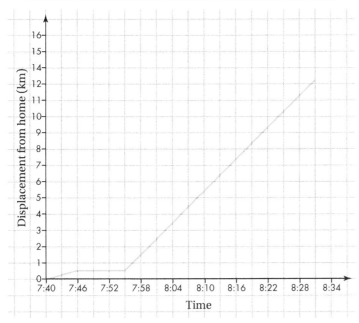

b Since school starts at 8:30 a.m., the boy arrives 1 minute late.

Velocity–time graphs

The velocity–time graph below shows the motion of a car.

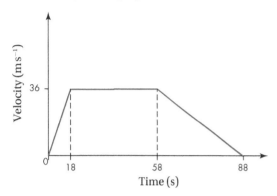

There are three stages to the journey:

> during the first stage, the car accelerates

> during the second stage, the car drives at a steady speed

> during the third and final stage, the car decelerates to rest.

You are now going to be shown how the gradient and area under the graph give the same results for acceleration and distance travelled as when you use the equations of uniformly accelerated motion.

For the first stage of the journey, the car travels at a constant acceleration from rest to a velocity of $36\,\text{m}\,\text{s}^{-1}$ in 18 seconds.

Using $v = u + at$, you can find the acceleration.

$$a = \frac{v - u}{t} = \frac{36 - 0}{18} = 2\,\text{m}\,\text{s}^{-2}$$

Note that this is the same as the gradient of the line segment joining $(0, 0)$ and $(18, 36)$.

Using $s = \frac{1}{2}(u + v)t$, you can find the displacement.

$$s = \frac{1}{2}(u + v)t = \left(\frac{36 + 0}{2}\right) \times 18$$

$$= \frac{1}{2} \times 18 \times 36$$

$$= 324\,\text{m}$$

Note that this is the same as the area of the triangle with a base of 18 and a height of 36.

For the second stage of the journey, the car travels at a steady speed of $36\,\text{m}\,\text{s}^{-1}$ for 40 seconds. Note that both the acceleration and the gradient are zero.

Given that $a = 0$, you can rewrite $s = ut + \frac{1}{2}at^2$ as $s = ut$.

$s = 36 \times 40$

$\quad = 1440\,\text{m}$

$s = ut$ is equivalent to saying that for a constant speed distance = speed × time

Note that this is the same as the area of the rectangle with a base of 40 and a height of 36.

For the final stage of the journey, the car travels at a constant deceleration from $36\,\text{m s}^{-1}$ to rest in 30 seconds.

Using $v = u + at$, you can find the deceleration.

$a = \dfrac{v - u}{t} = \dfrac{0 - 36}{30} = -1.2\,\text{m s}^{-2}.$

Note that this is the same as the gradient of the line segment joining $(58, 36)$ and $(88, 0)$.

Using $s = \frac{1}{2}(u + v)t$, you can find the displacement.

$s = \frac{1}{2}(u + v)t = \left(\dfrac{0 + 36}{2}\right) \times 30$

$\quad = \frac{1}{2} \times 30 \times 36$

$\quad = 540\,\text{m}$

Note that this is the same as the area of the triangle with a base of 30 and a height of 36.

Stop and think How can you find the total distance from the whole graph without splitting it into two triangles and a rectangle?

Example 10

A car accelerates from rest at $1.6\,\text{m s}^{-2}$ for 25 seconds. It then maintains a steady speed for 42 seconds before decelerating to rest in 20 seconds.

a Draw a velocity–time graph for this journey.

b Find the deceleration of the car as it comes to rest.

c Find the total distance travelled.

Solution

a There are three stages to this journey.

For the first stage, you are told that the car starts at rest (so $u = 0\,\text{m s}^{-1}$) and accelerates at $1.6\,\text{m s}^{-2}$ (so $a = 1.6\,\text{m s}^{-2}$) for 25 seconds (so $t = 25\,\text{s}$).

Substituting into $v = u + at$.

$v = u + at = 0 + 1.6 \times 25 = 40\,\text{m s}^{-1}$

For the second stage, the car drives at a steady $40\,\mathrm{m\,s^{-1}}$ for 42 seconds, from 25 seconds to 67 seconds.

For the third stage, the car decelerates to rest in 20 s, from 67 s to 87 s.

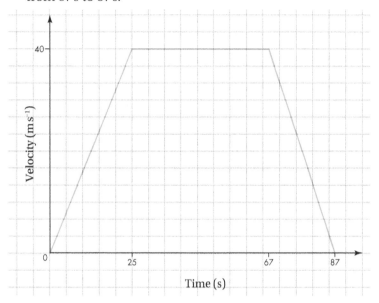

b For the third stage, the car has $u = 40\,\mathrm{m\,s^{-1}}$, $v = 0\,\mathrm{m\,s^{-1}}$ and $t = 20\,\mathrm{s}$, from which

$$a = \frac{v - u}{t} = \frac{0 - 40}{20} = -2\,\mathrm{m\,s^{-2}}.$$

> The acceleration is $-2\,\mathrm{m\,s^{-2}}$ but the deceleration is $2\,\mathrm{m\,s^{-2}}$. To use the constant acceleration formulae, you always need to use the acceleration value.

c The total distance travelled is given by the area under the graph. In this example, the graph is in the shape of a trapezium.

The total journey time is given by $25 + 42 + 20 = 87\,\mathrm{s}$.

The area is given by

$$\text{area} = \frac{1}{2}(a + b)h.$$

$$\frac{1}{2} \times (87 + 42) \times 40 = 2580$$

Total distance travelled = 2580 m.

Example 11

A bus departs from a bus stop at a constant acceleration for 15 seconds until it reaches a speed of $20\,\mathrm{m\,s^{-1}}$. It then drives at $20\,\mathrm{m\,s^{-1}}$ until it is X seconds from its destination, at which point it decelerates at $0.8\,\mathrm{m\,s^{-2}}$, coming to rest at the next bus stop. The bus stops are 1100 m apart and the total journey between the bus stops takes T seconds.

a Draw a velocity–time graph.

b Calculate the value of X.

c Calculate the value of T.

Solution

a There are three stages to this journey.

For the first part, the bus accelerates from $0\,\text{m}\,\text{s}^{-1}$ to $20\,\text{m}\,\text{s}^{-1}$ in 15 seconds.

For the second part, the bus drives at $20\,\text{m}\,\text{s}^{-1}$ until it is $(T - X)$ seconds from its destination.

For the third part, the bus decelerates to rest in X seconds.

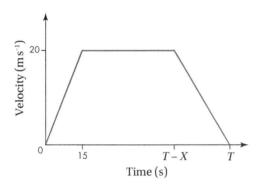

b When the bus decelerates, you have $a = -0.8\,\text{m}\,\text{s}^{-2}$, $v = 20\,\text{m}\,\text{s}^{-1}$, $u = 0\,\text{m}\,\text{s}^{-1}$ and $t = X\,\text{s}$.

$v = u + at$

$t = \dfrac{v - u}{a}$

$X = \dfrac{0 - 20}{-0.8}$

$X = 25$ seconds

c The total distance travelled is the area under the graph.

The second part of the journey takes $[(T - X) - 15]$ s, which is $(T - 40)$ s, since $X = 25$.

$\text{Area} = \dfrac{1}{2}(a + b)h$

$1100 = \dfrac{1}{2}(T + T - 40) \times 20$

$\qquad = 10(2T - 40)$

$110 = 2T - 40$

$150 = 2T$

$\quad T = 75$ seconds

> The distance $1100\,\text{m}$ is represented by the area under the graph.
>
> The parallel sides of the trapezium are T and $(T - 40)$.

Exercise 2.4A

1

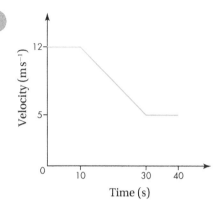

The velocity–time graph shows the journey of a cyclist. Initially the cyclist was travelling at a steady speed of $12\,\text{m}\,\text{s}^{-1}$ for 10 seconds. He then decelerated for 20 seconds until he was travelling at $5\,\text{m}\,\text{s}^{-1}$, which he then maintained for the final 10 seconds.

a Find the cyclist's deceleration.

b Find the total distance travelled by the cyclist.

2 A coach sets off from a town at 11:20 a.m. and travels at $45\,\text{km}\,\text{h}^{-1}$ for 40 minutes. It stays at its destination for an hour, then returns to the town at $50\,\text{km}\,\text{h}^{-1}$.

a Sketch a displacement–time graph to show the motion of the coach.

b At what time does the coach return to the town?

3 A particle moves at a steady velocity of $3\,\text{m}\,\text{s}^{-1}$ for 12 seconds followed immediately by a steady velocity of $-2\,\text{m}\,\text{s}^{-1}$ for 15 seconds.

a Sketch a displacement–time graph to show the displacement of the particle from its initial position.

A second particle starts moving from the same initial position three seconds after the first particle starts to move. This particle moves at a steady velocity of $1.6\,\text{m}\,\text{s}^{-1}$.

b Sketch the motion of the second particle on the same displacement–time graph.

c At what time and displacement do the particles meet?

4 Two cars, A and B, are travelling in opposite directions along a straight road between two towns 42 km apart. Both cars leave at 4:45 p.m. Car A travels at $75\,\text{km}\,\text{h}^{-1}$ for 20 minutes, stops for 10 minutes and then completes its journey at $51\,\text{km}\,\text{h}^{-1}$. Car B travels at $72\,\text{km}\,\text{h}^{-1}$ until 4:55 p.m., stops for 20 minutes and then completes its journey at $72\,\text{km}\,\text{h}^{-1}$.

a Sketch a displacement–time graph to show the motion of the cars.

b At what time are the cars in the same place?

5 A car accelerates from rest for 30 s at a constant acceleration, stays at a steady speed of $V\,\text{m}\,\text{s}^{-1}$ for X s, then decelerates at a constant acceleration, coming to a standstill 100 s after it set off. It travels a quarter of the total distance during the first 30 s.

a Sketch a velocity–time graph to show the motion of the car.

b Find the value of X.

c Given that the car decelerates at $0.9\,\text{m}\,\text{s}^{-2}$, find the value of V.

PS **6** A motorcyclist passes a police car at a steady $18\,\text{m}\,\text{s}^{-1}$. Ten seconds later the police car gives chase, accelerating to $24\,\text{m}\,\text{s}^{-1}$ in $10\,\text{s}$, then pursuing the motorcyclist at this steady speed.

    a Represent this information from the time the motorcycle passes the car in a velocity–time diagram.

    b How long after the motorcyclist passed the police car did it take for the police car to catch up?

    c How far did the police car travel in pursuit?

PS **7** Two particles, P and Q, start moving along the same straight line at the same time. P starts at O and moves at $1.25\,\text{m}\,\text{s}^{-2}$ for T seconds until it is D m from O. Q starts at a displacement of D m from O and moves at $-0.5\,\text{m}\,\text{s}^{-2}$ for T seconds until it is a displacement of $30\,\text{m}$ from O.

    a Sketch a displacement–time graph to show the motion of the particles.

    b Find the value of T.

    c Find the value of D.

PS **8** Two athletes compete in a race. Both start from rest. The first athlete initially runs a constant acceleration of $0.8\,\text{m}\,\text{s}^{-2}$ for $10\,\text{s}$, then stays at a constant speed until the end of the race. She takes a total of time of $105\,\text{s}$ to complete the race. The second athlete initially runs at a constant acceleration of $0.225\,\text{m}\,\text{s}^{-2}$ for $40\,\text{s}$, then reduces to a different constant acceleration which she maintains until the end of the race. The second athlete wins the race, finishing three seconds sooner than her opponent.

    a Sketch a velocity–time graph to show the motion of the two athletes.

    b Find the length of the race (in m).

    c Find the speed of the winning athlete as she finishes the race.

9 Show, for a journey of total time T, which involves a constant acceleration from rest for T_1 seconds to $V\,\text{m}\,\text{s}^{-1}$, a steady speed of $V\,\text{m}\,\text{s}^{-1}$ for T_2 seconds and a final constant deceleration to rest for T_3 seconds, that the distance given by the equations of uniformly accelerated motion can be written as the area of the trapezium with parallel sides of T and T_2 and a perpendicular height of V.

PS **10** Two vehicles start and finish at the same time and the same place. Both vehicles start from and finish at rest. Vehicle A accelerates at $1.5\,\text{m}\,\text{s}^{-2}$ for $20\,\text{s}$, then maintains a steady speed for $1\,\text{min}\,10\,\text{s}$ before decelerating. Vehicle B accelerates at $0.8\,\text{m}\,\text{s}^{-2}$ for $50\,\text{s}$, then maintains a steady speed for $20\,\text{s}$ before decelerating.

    a Sketch a velocity–time graph to illustrate this situation.

    b Find the total time taken.

    c Find the total distance travelled.

    d Find the deceleration of each vehicle.

11 A skydiver leapt from a plane. She spent the first 45 seconds in free fall, then immediately deployed her parachute, descending to the ground at a steady speed of $8\,\text{m}\,\text{s}^{-1}$ for 3 minutes.

    a Draw a velocity–time graph for her descent.

    b Draw a distance–time graph for her descent.

   c Draw an acceleration–time graph for her descent.

   d Calculate the total descent of the skydive, in metres, given by this model.

   e Will the actual descent be longer or shorter than your answer to **part d**? Explain your answer.

 12 A van drives at a constant velocity, $V\,\text{m s}^{-1}$, for three-quarters of a minute, then decelerates at $1.6\,\text{m s}^{-2}$ to rest in T seconds. The van travels a total distance of 2.3 km.

   a Sketch a velocity–time graph to show the motion of the van.

   b Prove that $T^2 + 90T - 2875 = 0$.

   c Find the value of V.

2.5 Solving kinematics problems using differentiation and integration

The equations of uniformly accelerated motion only apply when the acceleration is constant. When the acceleration is a function of time, then you will need to use calculus instead.

Differentiation

Velocity is the rate of change of displacement. If you have an expression for the displacement in terms of time (t), **differentiate** it to find an expression for the velocity. Similarly, acceleration is the rate of change of velocity. If you have an expression for the velocity in terms of time (t), differentiate it to find an expression for the acceleration.

Integration

Because differentiation and **integration** are inverse processes, you can also find the velocity and displacement if you know the acceleration. If you integrate an expression for the acceleration you will get an expression for the velocity and if you integrate an expression for the velocity you will get an expression for the displacement.

However, you need to remember to include a **constant of integration** (c), which you can then calculate by substituting known values.

For example, suppose that $a = 2t - 4$ and you know that the velocity is 3 ms^{-1} after 2 seconds.

Integrate the expression for a with respect to t to find an expression for the velocity v.

$$v = \int a\,dt \ = \ \int (2t - 4)\,dt \ = t^2 - 4t + c$$

Substitute $t = 2$ and $v = 3$ to find the value of the constant of integration.

$3 = 2^2 - 4 \times 2 + c$

$3 = 4 - 8 + c$

$c = 7$

Hence, $v = t^2 - 4t + 7$.

In summary:

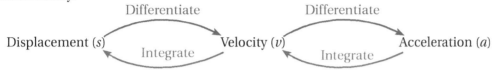

Differentiate

Differentiate

Displacement (s)   Integrate   Velocity (v)   Integrate   Acceleration (a)

Example 12

After t seconds, the displacement of a particle is given by $s = t^3 - 27t^2 + 246t + 16$, where s is in metres.

Find the acceleration of the particle at each time that the velocity is $15\,\text{ms}^{-1}$.

Solution

Find an expression for v by differentiating the expression for s with respect to t.

$$v = \frac{ds}{dt} = 3t^2 - 54t + 246$$

Find an expression for a by differentiating the expression for v with respect to t.

$$a = \frac{dv}{dt} = 6t - 54$$

KEY INFORMATION

Differentiate s to get v and differentiate v to get a.

Find the times at which the velocity is $15\,\text{ms}^{-1}$ by putting the expression for v equal to 15.

$3t^2 - 54t + 246 = 15$

Since this is a quadratic equation, subtract 15 from each side so that you can write it in the form $at^2 + bt + c = 0$.

$3t^2 - 54t + 231 = 0$

Simplify the equation by dividing through by the common factor of 3.

$t^2 - 18t + 77 = 0$

Factorise and solve.

$(t - 7)(t - 11) = 0$

$t = 7$ seconds or 11 seconds

Find the acceleration at each of the times by substituting for t.

When $t = 7$, $a = 6 \times 7 - 54 = -12\,\text{ms}^{-2}$.

When $t = 11$, $a = 6 \times 11 - 54 = 12\,\text{ms}^{-2}$.

Example 13

After t seconds, the acceleration of a particle is given by $a = (36 - 12t)\,\text{m s}^{-2}$.

After 5 seconds, the displacement of the particle is $11\,\text{m}$ and the velocity is $2\,\text{m s}^{-1}$.

Find the displacement of the particle when the particle is travelling at its maximum velocity.

Solution

Find an expression for v by integrating the expression for a with respect to t.

$$v = \int a\,\mathrm{d}t = \int (36 - 12t)\ \mathrm{d}t = 36t - 6t^2 + c$$

Substitute $t = 5$ and $v = 2$ to find the value of the constant of integration.

$$2 = 36 \times 5 - 6 \times 5^2 + c$$
$$= 180 - 150 + c$$
$$c = -28$$

Hence

$$v = 36t - 6t^2 - 28$$

Find an expression for s by integrating the expression for v with respect to t.

$$s = \int v\ dt = \int (36t - 6t^2 - 28)\ dt = 18t^2 - 2t^3 - 28t + c_2$$

Substitute $t = 5$ and $s = 11$ to find the value of the new constant of integration.

$$11 = 18 \times 5^2 - 2 \times 5^3 - 28 \times 5 + c_2$$
$$= 450 - 250 - 140 + c_2$$
$$c_2 = -49$$

Hence

$$s = 18t^2 - 2t^3 - 28t - 29$$

At maximum velocity, $a = 0\,\text{m s}^{-2}$. Solve the expression for a equal to zero.

$$36 - 12t = 0$$
$$t = 3\,\text{s}$$

Substitute $t = 3$ into the expression for s.

$$s = 18 \times 3^2 - 2 \times 3^3 - 28 \times 3 - 49 = -25\,\text{m}$$

> **KEY INFORMATION**
>
> Integrate a to get v and integrate v to get s. Remember to include a constant of integration.

> **KEY INFORMATION**
>
> Use a different constant of integration when integrating for a second time.

Example 14

A particle P moves in a straight line such that, after t seconds, its velocity, $v\,\mathrm{m\,s}^{-1}$, is given by

$$v = \begin{cases} t^2 + 2 & 0 \leqslant t \leqslant 4 \\ 2t^{\frac{3}{2}} + \frac{1}{2}t & t > 4. \end{cases}$$

Initially the particle is displaced 3 m from O.

a Find the acceleration of the particle when:

   i $t = 2$ **ii** $t = 9$.

b Find the total distance travelled during the first 9 seconds, correct to the nearest metre.

Solution

a **i** For $0 \leqslant t \leqslant 4$, differentiate $v = t^2 + 2$.

$$a = \frac{\mathrm{d}v}{\mathrm{d}t} = 2t$$

Substitute $t = 2$.

$$a = 2 \times 2 = 4\,\mathrm{m\,s}^{-2}$$

  ii For $t > 4$, differentiate $v = 2t^{\frac{3}{2}} + \frac{1}{2}t$.

$$a = \frac{\mathrm{d}v}{\mathrm{d}t} = 2 \times \frac{3}{2}t^{\frac{1}{2}} + \frac{1}{2} = 3t^{\frac{1}{2}} + \frac{1}{2}$$

Substitute $t = 9$ into this expression.

$$a = 3 \times 9^{\frac{1}{2}} + \frac{1}{2} = 3 \times 3 + \frac{1}{2} = 9\frac{1}{2}\,\mathrm{m\,s}^{-2}$$

b For $0 \leqslant t \leqslant 4$, integrate $v = (t^2 + 2)$.

$$s = \int v\ \mathrm{d}t = \int (t^2 + 2)\ \mathrm{d}t = \frac{1}{3}t^3 + 2t + c$$

Substitute $t = 0$ and $s = 3$.

$$3 = \frac{1}{3} \times 0^3 + 2 \times 0 + c$$

$$= c$$

Hence

$$s = \frac{1}{3}t^3 + 2t + 3.$$

Substitute $t = 4$ into this expression.

$$s = \frac{1}{3} \times 4^3 + 2 \times 4 + 3 = \frac{97}{3}$$

For $t > 4$, integrate $2t^{\frac{3}{2}} + \frac{1}{2}t$.

$$s = \int v\ \mathrm{d}t = \int (2t^{\frac{3}{2}} + \frac{1}{2}t)\ \mathrm{d}t = \frac{4}{5}t^{\frac{5}{2}} + \frac{1}{4}t^2 + c_2$$

Find s when $t = 4$ so that you can use it to find the constant of integration for the second part.

Substitute $t = 4$ and $s = \dfrac{97}{3}$.

$$\frac{97}{3} = \frac{4}{5} \times 4^{\frac{5}{2}} + \frac{1}{4} \times 4^2 + c_2$$

$$= \frac{128}{5} + 4 + c_2$$

$$c_2 = \frac{41}{15}$$

$$s = \frac{4}{5}t^{\frac{5}{2}} + \frac{1}{4}t^2 + \frac{41}{15}$$

When $t = 9$, $s = \dfrac{4}{5} \times 9^{\frac{5}{2}} + \dfrac{1}{4} \times 9^2 + \dfrac{41}{15} = 217\,\text{m}$.

Exercise 2.5A

1 The displacement, in metres, of a particle is given by $s = 13t + 20 - 4t^2$, where t is the time in seconds.

   a Find the velocity of the particle when $t = 2\,\text{s}$.

   b Show that the acceleration is constant.

2 After t seconds, a particle has a displacement of $s = (5t^3 - 7t^2)$ metres.

   a Find expressions for the velocity and acceleration in terms of t.

   b Find the velocity after 2 seconds.

   c Find the acceleration after 3 seconds.

3 The velocity of a particle after t seconds is given by $v = (t - 2)\,\text{m s}^{-1}$. After 6 seconds, the displacement is $11\,\text{m}$ from O.

   a Find an expression for the displacement, s.

   b Find the displacement after 8 seconds.

   c Find the velocity when the particle has a displacement of $53\,\text{m}$ from O.

4 At time t s, a particle is $(t^3 - 8t^2 + 5t + 10)\,\text{m}$ from the origin.

   a At what times is the particle instantaneously at rest?

   b At what times is the velocity positive?

   c At what times is the acceleration positive?

   d Find the velocity and displacement when the particle is accelerating at $2\,\text{m s}^{-2}$.

5 The velocity, in m s^{-1}, of a particle is given by $v = 18t - t^2$.

   Find the acceleration at both instants when the particle has a velocity of $65\,\text{m s}^{-1}$.

6 The displacement of a particle after t seconds is given by $s = 8\sqrt{t} - \dfrac{64}{t}$.

   a Show that the particle is at the origin when $t = 4$.

   b Find the velocity and acceleration of the particle when $t = 4$.

7 The acceleration of a particle after t seconds is given by $a = (6t - 20)\,\text{ms}^{-2}$. After 12 seconds, the velocity is $223\,\text{ms}^{-1}$.

   a Find an expression for the velocity, v.

   After 7 seconds, the displacement is $40\,\text{m}$.

   b Find an expression for the displacement, s.

   c Find the times at which the particle is at the origin.

   d Find the times at which the particle is at travelling at $19\,\text{ms}^{-1}$.

8 A particle has a displacement, in cm, of $s = (t^4 - 8t^3 + 22t^2 - 20t + q)$ from O after t seconds. After 4 seconds, the particle is $33\,\text{cm}$ from O.

   a Find the value of q.

   b Find the displacement of the particle each time the particle is moving at $4\,\text{cm s}^{-1}$.

   c Find the displacement of the particle from O when the particle is accelerating at $188\,\text{cm s}^{-2}$.

9 A particle has an acceleration of $(6t - 30)\,\text{ms}^{-2}$. Initially, the particle is at O with a velocity of $72\,\text{ms}^{-1}$.

   a Find the minimum velocity of the particle.

   b When is the particle at instantaneous rest?

   c Find the total distance travelled during the first ten seconds.

10 A particle has an acceleration of $(9 - 2t)\,\text{ms}^{-2}$. Initially, the particle is at rest at the origin. Find the distance travelled by the particle:

   a during the first two seconds

   b during the fifth second

   c during the ninth and tenth seconds.

11 A particle P moves in a straight line such that, after t seconds, its acceleration, $a\,\text{ms}^{-2}$, is given by

$$a = \begin{cases} \dfrac{1}{2}(t + 6) & 0 \leqslant t \leqslant 4 \\ \dfrac{320}{t^3} & 4 < t \leqslant 8 \end{cases}$$

   Initially the particle is at the origin travelling at a velocity of $2\,\text{ms}^{-1}$.

   a Find the speed of the particle when:

     i $t = 3$ **ii** $t = 5$.

   b Find the total distance travelled during the 8 seconds.

12 A particle P moves in a straight line such that, after t seconds, its velocity, $v\,\mathrm{m\,s^{-1}}$, is given by

$$v = \begin{cases} 2t + 4 & 0 \leqslant t \leqslant 6 \\ 20 - (t - 8)^2 & 6 < t \leqslant 10 \end{cases}$$

Initially, the particle is at the origin.

a Draw a velocity–time graph to illustrate the journey.

b Find the maximum speed of P.

c Find the acceleration of the particle when:

i $t = 5$ **ii** $t = 9$.

d Find the displacement at $t = 5$ and $t = 9$.

 13 A particle is initially at the origin, O. It travels at a constant acceleration from an initial velocity of $6\,\mathrm{m\,s^{-1}}$ to $33\,\mathrm{m\,s^{-1}}$ in 18 seconds. Subsequently, its acceleration is given by

$a = \frac{1}{4}(t - 12)\,\mathrm{m\,s^{-2}}$. Find the total distance travelled by the particle during the first 30 seconds.

Mathematics in life and work: Group discussion

There is an automatic fine issued to any railway company that does not meet prescribed targets. These targets include ensuring that a certain percentage of all trains running on a particular line are on time. For example, a fine might be incurred if arrival times of fewer than 95% of trains are not within three minutes of the time stated on the timetable.

In your role at a railway company, you are in charge of train timetables for a particular train line. You need to ensure that the timings are flexible enough to reduce the likelihood of incurring a fine but also efficient enough to encourage people to use the service rather than find an alternative means of transport.

Two of the stations on the line are coded as C and D, where CD is 2.5 km. Your task is to determine an appropriate length of time for the train to travel between C and D for recording on the timetable. You are expected to model the most appropriate speed–time graph and you may use straight line or polynomial curves.

1 If the train is to accelerate to its maximum speed as efficiently as possible, what shape might you use for the initial stage of the motion?

2 If the train is to decelerate as it arrives at D as efficiently as possible, what shape might you use for the final stage of the motion?

3 Devise a model which uses your answers for **questions 1** and **2** and has a total distance of 2.5 km.

SUMMARY OF KEY POINTS

> Distance and speed are scalar quantities. Displacement and velocity are vector quantities. Acceleration is a vector quantity but the same word is also used to describe the magnitude of the acceleration vector.

> Displacement, velocity, acceleration and time are related by the equations of uniformly accelerated motion:

> > $v = u + at$

> > $v^2 = u^2 + 2as$

> > $s = \frac{1}{2}(u + v)t$

> > $s = ut + \frac{1}{2}at^2$

> > $s = vt - \frac{1}{2}at^2$

> Make sure that you write the quantities in SI units (m and s). You will need to decide which direction to take as positive.

> When an object is travelling vertically, it experiences an acceleration of $10\,\mathrm{m\,s^{-2}}$ downwards due to gravity.

> For a displacement–time graph, the gradient is the velocity.

> For a velocity–time graph, the gradient is the acceleration and the area under the graph is the displacement.

> When displacement, velocity or acceleration is a function of time, use calculus. Remember to use a constant of integration when integrating.

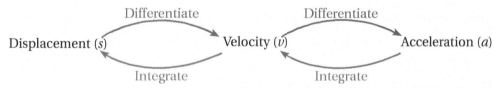

EXAM-STYLE QUESTIONS

1. A car moves with constant acceleration along a straight horizontal road. The car passes point A with speed $1.5\,\mathrm{m\,s^{-1}}$ and 6 s later it passes point B, where $AB = 90\,\mathrm{m}$.

   a Find the acceleration of the car.

   When the car passes the point C, it has speed $37.5\,\mathrm{m\,s^{-1}}$.

   b Find the distance AC.

2. A small ball is projected vertically upwards from a point A. The greatest height reached by the ball is 80 m above A. Calculate:

   a the speed of projection

   b the time between the instant that the ball is projected and the instant it returns to A.

PS **3** A particle has a displacement of $s = (t^3 - 14t^2 + 48t)$ metres, where t is the time in seconds.

 a At what times does the particle return to its original position?

 b Find the displacement of the particle when it is travelling at its minimum velocity.

PS **4** A particle is accelerating at $(pt - 10)\,\mathrm{m\,s^{-2}}$. Two seconds after it sets off it has a velocity of $30\,\mathrm{m\,s^{-1}}$. Half a second after that it has a velocity of $52\,\mathrm{m\,s^{-1}}$. If the particle is initially at the origin, find its position after three seconds.

5 A rocket is launched vertically upwards from a point 20 m above the ground at a velocity of $37.5\,\mathrm{m\,s^{-1}}$.

 a Show that when the rocket hits the ground $2t^2 - 15t - 8 = 0$.

 b Hence find the time at which the rocket hits the ground.

 c Find the total distance travelled by the rocket.

PS **6** Three posts P, Q and R, are fixed in that order at the side of a straight horizontal road. A car passes Q five seconds after passing P, and the car passes R ten seconds after passing Q. The distance from P to Q is 40 m and the distance from Q to R is 20 m. The car is moving along the road with constant acceleration $a\,\mathrm{m\,s^{-2}}$ and the speed of the car as it passes P is $u\,\mathrm{m\,s^{-1}}$.

 Find:

 a the value of u

 b the value of a.

7 A small ball is projected vertically upwards from ground level with speed $u\,\mathrm{m\,s^{-1}}$. The ball takes 6 s to return to ground level.

 a Draw a velocity–time graph to represent the motion of the ball during the first 6 s.

 b The maximum height of the ball above the ground during the first 6 s is 45 m. Find the value of u.

 c State a physical factor that has been ignored in this model.

PS **8** A particle moving along the x-axis is initially at the origin O. At time t seconds, the velocity of the particle is given by $v = (3t^2 - 12t + 11)\,\mathrm{m\,s^{-1}}$ in the positive x-direction. Find the distance of the particle from O when it is moving with minimum velocity.

9 Two cars, A and B, are moving in the same direction along a straight horizontal road. At time $t = 0$, they are side by side, passing a point O on the road. Car A travels at a constant speed of $36\,\mathrm{m\,s^{-1}}$. Car B passes O with a speed of $18\,\mathrm{m\,s^{-1}}$, and has constant acceleration of $3\,\mathrm{m\,s^{-2}}$.

 Find:

 a the speed of B when it has travelled 42 m from O

 b the distance from O of A when B is 42 m from O

 c the time when B overtakes A.

PS **10** A car starts from rest at a point S on a straight race track. The car moves with constant acceleration for 30 s, reaching a speed of $24\,\mathrm{m\,s^{-1}}$. The car then travels at a constant speed of $24\,\mathrm{m\,s^{-1}}$ for 150 s. Finally, it moves with constant deceleration, coming to rest at a point F.

a Sketch a velocity–time graph to illustrate the motion of the car.

The distance between S and F is 4.5 km.

b Calculate the total time the car takes to travel from S to F.

A motorcycle starts at S, 20 s after the car has left S. The motorcycle moves with constant acceleration from rest and passes the car at a point P which is 1920 km from S. When the motorcycle passes the car, the motorcycle is still accelerating and the car is moving at a constant speed. Calculate:

c the time the motorcycle takes to travel from S to P

d the speed of the motorcycle at P.

(PS) 11 A runner is competing in a race. Starting from rest, he accelerates at $4\,\mathrm{m\,s^{-2}}$ for 2 seconds, then at $\frac{1}{3}\,\mathrm{m\,s^{-2}}$ for 3 seconds and at $\frac{3}{13}\,\mathrm{m\,s^{-2}}$ for 13 seconds, before finally decelerating at $2.4\,\mathrm{m\,s^{-2}}$ until he comes to rest at the finish line. What was the runner's average speed for the race, correct to 3 significant figures?

(PS) 12 A particle P moves along a straight line. After t seconds, the velocity of v is given by

$$v = \begin{cases} 8t - t^2 & 0 \leqslant t \leqslant 6 \\ 15 - \frac{1}{2}t & t > 6 \end{cases}$$

The particle is initially at the origin O.

Find:

a the acceleration of P when $t = 5$

b the total distance travelled by P during the first 40 seconds.

(PS) 13 A car moves along a horizontal straight road, passing two points A and B. At A, the speed of the car is $18\,\mathrm{m\,s^{-1}}$. When the driver passes A, she sees a warning sign, W, ahead of her, 180 m away. She immediately applies the brakes and the car decelerates with uniform deceleration, reaching W with speed $6\,\mathrm{m\,s^{-1}}$. At W, the driver sees that the road is clear. She then immediately accelerates the car with uniform acceleration for 18 s to reach a speed of $V\,\mathrm{m\,s^{-1}}$ ($V > 18$). She then maintains the car at a constant speed of $V\,\mathrm{m\,s^{-1}}$. Moving at this constant speed, the car passes B after a further 33 s.

a Sketch a velocity–time graph to illustrate the motion of the car as it moves from A to B.

b Find the time taken for the car to move from A to B.

The distance from A to B is 1.2 km.

c Find the value of V.

(PS) 14 A particle P moves on the x-axis. The acceleration of P at time t seconds is $(t - 6)\,\mathrm{m\,s^{-2}}$ in the positive x-direction. The velocity of P at time t seconds is $v\,\mathrm{m\,s^{-1}}$. When $t = 0$, $v = 10$. Find:

a v in terms of t

b the values of t when P is instantaneously at rest

c the distance between the two points at which P is instantaneously at rest.

PS **15** A ball is projected vertically upwards at $35\,\text{m}\,\text{s}^{-1}$ from the top of a building. 2 s later, a second ball is projected vertically upwards at $24\,\text{m}\,\text{s}^{-1}$, from the same place. The balls land on the ground at the same time. Find the height of the building.

PS **16** A particle travels in a straight line with a constant acceleration through points A, B and C. BC is three times the distance of AB. The particle passes B 3 s after it passes A and passes C 7 s after it passes B. Given that the velocity of the particle at B is $4.5\,\text{m}\,\text{s}^{-1}$ faster than at A, find the length AC.

MM **17** The displacement of a particle, P, from the origin after t s is given by $s = t^2(t + k)$, $k \neq 0$.

Given that the particle comes to instantaneous rest after 6 s, show that:

a the acceleration of the particle after 13 s is $60\,\text{m}\,\text{s}^{-2}$

b the particle travels 316 m during the first 10 s.

Mathematics in life and work

Your role at a bus company means that you are responsible for updating the timetable for a particular route to make it more reliable. In the previous model of the route between two bus stops labelled X and Y, a bus left X and accelerated uniformly from rest at $2\,\text{m}\,\text{s}^{-2}$ until it reached a speed of $12\,\text{m}\,\text{s}^{-1}$. The bus would then maintain this speed for 5 seconds before decelerating uniformly to rest at Y. The total journey took 20 s.

1 Find the deceleration of the bus.

2 Find the distance between X and Y.

You have devised a new model for the motion of the bus. Instead of accelerating uniformly for the first 6 s, the velocity of the bus is defined as $v = \frac{1}{3}t(12 - t)$. Once the bus reaches a speed of $12\,\text{m}\,\text{s}^{-1}$, it immediately decelerates uniformly in the same way as the previous model.

3 Find the total journey time for your new model.

3 MOMENTUM

Mathematics in life and work

When a rocket is sent into space it is easy to assume that it is pushed off the launchpad by the fuel blasting out of its exhaust. However, if this were the case then the rocket would no longer be propelled once it had moved off the launchpad – and once in the vacuum of space it would be unable to move at all. In fact, the rocket moves due to conservation of momentum.

When two vehicles are involved in a road crash the impact does not always affect them both in the same way. If a truck collides with a small car, you might expect the car to come off worse because the truck will have a greater momentum. In this chapter, you will look into the mathematics behind this type of situation.

The mathematics of momentum can be useful in many different careers – for example:

› If you were designing a car, you would include a feature known as a 'crumple zone'. This is an area which allows the momentum of a car to be displaced safely, so that the driver and passengers are kept safe. Airbags are another factor to consider with momentum. This time, it is the momentum of the person which needs to be displaced safely, allowing them to slow down less rapidly than they would have done in a collision.

› The design of sports equipment is always cutting edge. If you were designing a bat for baseball or cricket, you would factor in how the weight and speed of the bat affect the momentum of the ball.

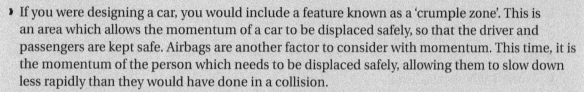

› In the train industry, it is vital to understand the speed at which trains can meet in order to be coupled together, or the maximum safe speed at which buffers will be effective. Railway engineers understand how the weight of the train will affect its stopping distance as it approaches a station, and therefore depend on mathematics of momentum when calculating top speeds.

› If your job included designing projectile machines – for example, as a manufacturer of a tennis machine that shoots out balls as practice – momentum is an important element in understanding how the projectile moves and the recoil of the equipment shooting the projectile.

LEARNING OBJECTIVES

You will learn how to:

› use the definition of linear momentum and show understanding of its vector nature

> use conservation of linear momentum to solve problems which may be modelled as the direct impact of two bodies

> understand the case where the bodies coalesce on impact.

LANGUAGE OF MATHEMATICS

Key words and phrases you will meet in this chapter:

coalesce, collide, conservation of linear momentum, mass, momentum

PREREQUISITE KNOWLEDGE

You should already know:

> the difference between the speed and velocity of an object

> how to solve a basic linear equation

> how to convert between different units of mass and velocity.

You should be able to complete the following questions correctly:

1 Make x the subject of these formulae.

    a $y = 3x + b$ **b** $5t + 4x = mx$ **c** $\dfrac{g}{x + n} = y$

2 Fill in the blanks for these equalities.

    a $30\,\text{m}\,\text{s}^{-1}$ = _____ $\text{km}\,\text{h}^{-1}$

    b 15 km per min = _____ mm per second

    c $13\,\text{m}\,\text{s}^{-1} \approx$ _____ miles per hour

    d 100 g = _____ kg

    e 2 metric tonnes = _____ kg

3 Two cars are travelling in the same direction on an expressway. The first is moving at $120\,\text{km}\,\text{h}^{-1}$. The second is moving at $110\,\text{km}\,\text{h}^{-1}$. What is the velocity of the second car relative to the first car?

3.1 Calculating the momentum

The **momentum**, p**,** of a particle is the product of its **mass**, m with its velocity, v:

$p = mv$.

So any object that is moving will have momentum. The faster the object is moving or the greater its mass, the greater the momentum.

In this chapter, we will only consider motion in one dimension. This means that the velocity will only be acting in one line, so the direction will either be forwards or backwards. Although a velocity can be represented with a vector, you will only need to use a single number to state the velocity. However, the velocity can be negative, so it is important to know clearly which direction is taken to be *positive*, often indicated with an arrow.

> **KEY INFORMATION**
>
> momentum =
> mass (in kg) × velocity (m s^{-1})
>
> The standard unit of momentum is kg m s^{-1}.

If two cars are travelling towards each other at the same speed, how are their *velocities* the same, and how are they different? If a car is travelling at a constant speed around a bend, does the momentum change?

Example 1

A car of mass 1200 kg is travelling at a velocity of 45 km h^{-1}. Calculate the momentum of the car, giving your answer in kg m s^{-1}.

Solution

$45\,\text{km h}^{-1} = 45\,000\ \text{m h}^{-1} = \dfrac{45\,000}{60 \times 60}\ \text{m s}^{-1} = 12.5\,\text{m s}^{-1}$

Unless stated otherwise, convert to standard units – in this case, m s^{-1}.

So $p = 1200 \times 12.5 = 15\,000\,\text{kg m s}^{-1}$.

Under what circumstances could you say a car has a negative velocity? Can an object ever have a negative mass?

Example 2

A cart has a momentum of 12 kg m s^{-1}. Write an expression for the new momentum under the following circumstances.

a A load is added to the cart, doubling the total mass.

b Load is added, such that the mass is multiplied by a factor of n.

c The mass is multiplied by a factor of n, and the speed is divided by a factor of m.

Solution

a $p = m \times v$

If the mass doubles, so does the momentum. The momentum of the cart is now $12 \times 2 = 24\,\text{kg m s}^{-1}$.

b The momentum is now $12 \times n$ or $12n\,\text{kg m s}^{-1}$.

c The momentum is now $\dfrac{12n}{m}\ \text{kg m s}^{-1}$.

Medieval soldiers sometimes used battering rams to break open castle gates. How might they increase the momentum of the battering ram to make it more effective? What limitations are there to your solution?

Exercise 3.1A

1 Calculate the momentum of:

  a a tennis ball of mass 0.056 kg and velocity 35 m s^{-1}

  b a piano of mass 160 kg falling with a speed of 10 m s^{-1}

  c a lion of mass 180 kg running at a speed of 30 km h^{-1}

  d a wrecking ball of mass 860 kg swung at 5 m s^{-1}

  e a man of mass 80 kg running at a pace of 10 km per hour.

2 Calculate the momentum of an electron of mass 9.1×10^{-31} kg that is moving at 2.2×10^6 m s^{-1}.

3 What is the speed of a train that has a momentum of 500 000 kg m s^{-1} and a mass of 100 metric tonnes?

4 A car of mass 1.3 metric tonnes is reversing down a driveway with a speed of 3 m s^{-1}. Calculate the momentum of the car, making the direction clear.

5 A bullet leaves the end of a gun at 860 m s^{-1} with a momentum of 38.7 kg m s^{-1}. What is the mass in grams of the bullet?

6 A boat has a mass of 120 000 kg and is travelling with a velocity of 3 m s^{-1} due north. Work out the momentum, p kg m s^{-1}, of the boat.

(C) 7 Estimate the momentum, p kg m s^{-1} of:

  a a cricket ball after being struck by a bat

  b a snowflake falling from a cloud

  c a small car on an expressway

  d a person running the 100 m race at the Olympics.

8 A sledge sliding along a frozen lake has momentum of p kg m s^{-1}. Write an expression for the new momentum, in terms of p under the following changes.

  a Half the mass and the same speed.

  b Twice the mass and the same speed.

  c Twice the mass and half the speed.

  d A third of the mass and twice the speed.

  e Four times the mass and half the speed.

9 An object of constant mass 8 kg is doubling in velocity every second. If it has an initial velocity of 5 m s^{-1}, after how many seconds is the momentum over 1000 kg m s^{-1}?

10 A car of mass 1.3 tonnes accelerates from a speed of 30 km h^{-1} at a rate of 3 m s^{-2} for 5 seconds.

  a What is the final momentum of the car?

  Given that the car has an initial speed of u m s^{-1}:

  b write an expression for the final momentum of the car.

11 Write an expression for the change in momentum of an object of mass m kilograms that has been uniformly accelerating for t seconds for a distance of s metres.

(C) Communication   Mathematical modelling   (PS) Problem solving

3.2 Conservation of linear momentum

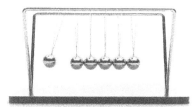

This picture shows a popular office toy called a Newton's cradle. It is set in action when one of the outer metal balls is pulled outwards and released. The ball then **collides** with the next ball in the cradle, which does not appear to move – however, the outer ball on the opposite side of the cradle will jump into the air. The balls must be of the same mass and hang so that there is no gap between them. Newton's cradle illustrates the principle of **conservation of linear momentum**.

Stop and think

What would happen if you lifted and dropped two of the balls on the left-hand side? What if you lifted and dropped three of them?

What would happen if you lifted both of the outer balls to the same height and released them at the same time?

If momentum is conserved, why doesn't a Newton's cradle continue moving forever?

Suppose that two particles A and B of mass m_A and m_B collide. What happens to the momentum of the particles? The two particles will collide, which will cause a change of velocity in both particles. Let their initial velocities be u_A and u_B and their final velocities be v_A and v_B.

We can say that

$$m_A u_A + m_B u_B = m_A v_A + m_B v_B$$

This result is known as the principle of conservation of (linear) momentum.

KEY INFORMATION

When there are no external forces on a system, the total momentum remains constant. In other words, the total momentum before a collision is equal to the total momentum after a collision.

Example 3

Two carts A and B, of masses 1.2 kg and 2.4 kg, respectively, are travelling in the same direction along a horizontal surface with initial velocities 9 m s^{-1} and 3 m s^{-1}. They directly collide, causing cart A to slow down to a velocity of 1 m s^{-1}. Using conservation of momentum, calculate the velocity of cart B immediately after the collision.

'Directly collide' means that they hit in the same direction as each others' motion, not at an angle, so this is a one-dimensional problem. If the objects are moving in the same line but opposite directions it is often called a 'head-on' collision.

Solution

Start by drawing a diagram to represent this information:

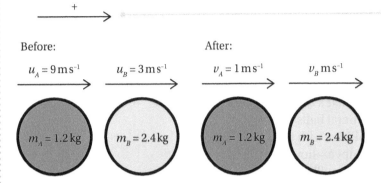

Note the arrow above the diagrams which defines the direction of a positive velocity.

Before:

$u_A = 9\,\text{m s}^{-1}$ $u_B = 3\,\text{m s}^{-1}$

$m_A = 1.2\,\text{kg}$ $m_B = 2.4\,\text{kg}$

After:

$v_A = 1\,\text{m s}^{-1}$ $v_B\,\text{m s}^{-1}$

$m_A = 1.2\,\text{kg}$ $m_B = 2.4\,\text{kg}$

$$m_A u_A + m_B u_B = m_A v_A + m_B v_B$$

Substitute in the known values.

All the units are consistent, so you do not need to state them in the calculations.

$$1.2 \times 9 + 2.4 \times 3 = 1.2 \times 1 + 2.4 \times v_B$$

Simplify.

$$10.8 + 7.2 = 1.2 + 2.4v_B$$

$$18 = 1.2 + 2.4v_B$$

$$16.8 = 2.4v_B$$

Divide both sides by 2.4.

$$7 = v_B$$

So the final velocity of cart $B = 7\,\text{m s}^{-1}$.

Stop and think   How might the solution change if you were told that carts A and B had an initial *speed* of $9\,\text{m s}^{-1}$ and $3\,\text{m s}^{-1}$?

Example 4

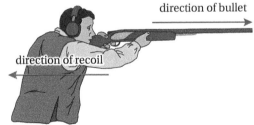

direction of bullet

direction of recoil

When a gun is fired, the recoil is the backward *momentum* caused when the bullet leaves the gun. Calculate the recoil, $x\,\text{m s}^{-1}$, of a gun of empty mass 1.5 kg when a bullet of mass 0.05 kg is fired at a speed of $255\,\text{m s}^{-1}$.

Solution

Draw a simple diagram of the situation.

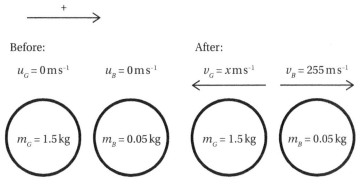

> The negative sign in front of the x is due to the fact that the recoil is in the opposite direction to the velocity of the bullet. The diagram has also made this clear.

$$m_A u_A + m_B u_B = m_A v_A + m_B v_B$$

The initial momentum is zero, so the final momentum must also sum to zero.

$$1.5 \times (-x) + 0.05 \times 255 = 0$$

$$-1.5x + 12.75 = 0$$

$$-1.5x = -12.75$$

$$x = 8.5\,\text{m s}^{-1}$$

> If it is unclear which direction a final velocity is in, you can simply choose it to be forwards or backwards. If you are correct, your numerical answer will be positive.

Stop and think   What are the main assumptions in **Example 4**? How would the model be changed if the gun was fixed to a stand? How might the conservation of momentum be used to reduce the recoil of a gun?

Mathematics in life and work: Group discussion

Safety features in cars have improved a great deal in recent years. Car designers need to ensure that any new car will be safe to drive and will protect the drivers and passengers. If a car is involved in a crash, the faster it is travelling the more likely it is to cause harm. Using conservation of momentum you can study the effects of the mass of a car involved in a crash.

1 What is the momentum of a car of mass 1500 kg and velocity 18 m s^{-1}?

Two cars are involved in a head-on collision: a small car, S, of mass 1000 kg and a large car, L, of mass 2500 kg. As a result of the collision, the large car is brought to a standstill. To compare the effect of the mass of different cars, assume that they are travelling at the same speed, but in opposite directions.

2 If the initial speeds of the cars are 30 m s^{-1} (approximately 108 km h^{-1}), what is the velocity of the smaller car immediately after the collision?

3 Copy and complete this table:

Velocity of *S* before collision (m s⁻¹)	30	25	20	15	10	5
Velocity of *S* after collision (m s⁻¹)						
Change in velocity (m s⁻¹)						

4 What is the impact on the smaller car of reducing the speed of approach? If the speed before collision drops by 5 m s⁻¹ for each car, what is the change of the speed after the collision for the smaller car? In many countries there are crash barriers on the fastest roads – how does the speed after collision differ if the car crashes into a wall or barrier?

Exercise 3.2A

1 For each of these direct impacts, calculate the missing mass or velocity.

a

Before:

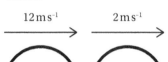

12 m s⁻¹   2 m s⁻¹

3 kg   2.8 kg

After:

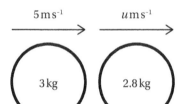

5 m s⁻¹   *u* m s⁻¹

3 kg   2.8 kg

b

Before:

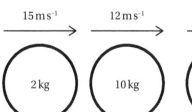

15 m s⁻¹   12 m s⁻¹

2 kg   10 kg

After:

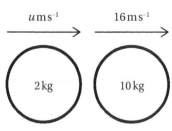

u m s⁻¹   16 m s⁻¹

2 kg   10 kg

c

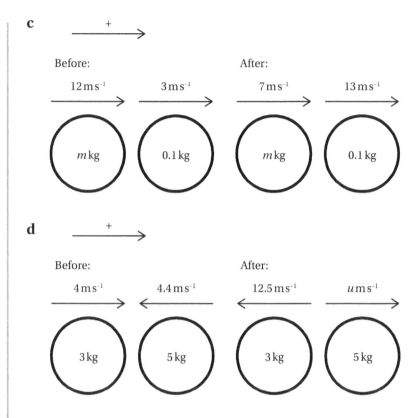

d

Before: After:

2 Teertha and her younger brother Kian are at the ice rink. They stand facing each other, then Teertha pushes against Kian and they both move backwards. Teertha estimates that she is moving at $1.5\,\text{m}\,\text{s}^{-1}$. Given that Teertha has mass $40\,\text{kg}$ and Kian has mass $30\,\text{kg}$, how fast is Kian moving?

3 In a theme park ride, a bumper car of mass $150\,\text{kg}$ is travelling at $3\,\text{m}\,\text{s}^{-1}$. It hits a stationary bumper car of mass $120\,\text{kg}$ head on, and the second bumper car then starts to move at $2\,\text{m}\,\text{s}^{-1}$. What is the subsequent velocity of the first car?

4 Two inflatable boats collide head on. The total mass (including the passengers) of boat A is $120\,\text{kg}$, and the total mass of boat B is $100\,\text{kg}$. Boat B was travelling toward boat A at $1.8\,\text{m}\,\text{s}^{-1}$, and boat A was travelling at $1.5\,\text{m}\,\text{s}^{-1}$ in the opposite direction. After the collision, boat A bounces backwards with a speed of $1.2\,\text{m}\,\text{s}^{-1}$. What are the new speed and direction of boat B after the collision?

PS 5 A snooker player hits the white ball, causing it to move at $0.8\,\text{m}\,\text{s}^{-1}$. The white ball directly hits a red ball, which has the same mass and size as the white. The red ball moves forward at $1.3\,\text{m}\,\text{s}^{-1}$ in the same line as the white ball. In which direction does the white ball move and at what speed?

6

When a naval cannon is fired it will move backwards, so it has a thick piece of rope which is designed to slow it down after firing. If a cannon has a mass of 100 kg, and is loaded with a cannonball of mass 5.5 kg, what is the initial speed of the cannon after the cannonball is fired at a horizontal speed of $140\,\mathrm{m\,s^{-1}}$?

(PS) 7 Two cars crash on a smooth, straight road. The report states that the cars are travelling at $5\,\mathrm{m\,s^{-1}}$ and $4\,\mathrm{m\,s^{-1}}$. It also notes that the cars have masses of 1200 kg and 1000 kg. The report does not state which car is which, or the initial directions of the cars. The cars collided, but did not stick together. How many different solutions are there for the subsequent movement of the two cars? The 1000 kg car was brought to a stop by the crash. Work out the new speed of the 1200 kg car in each case, making the direction clear.

8 Three snooker balls, A, B and C, of equal mass are spaced apart in a straight line. B and C are initially at rest. Ball A is hit such that it now has an initial velocity of $u\,\mathrm{m\,s^{-1}}$, moving towards B. After hitting B, A now moves with a *speed* of v_1. B moves at a *speed* of v_2 and it collides with C causing it to rebound with a *speed* of v_3. Ball C now moves at a speed of v_4. Given that ball B does not meet ball A for a second time, write an inequality for u in terms of the final velocities of balls A and C.

9 A light football of mass 200 g, initially at rest, starts rolling down a 3 m long slope. It accelerates at a rate of $6\,\mathrm{m\,s^{-2}}$. At the bottom of the hill it directly hits a standard football of mass 400 g, causing the light football to stop moving. How fast is the standard football moving?

10 Isabel and Gabriel are standing together on an ice rink. They push against each other and move such that it takes Gabriel 4 seconds to move 10 m. How long does it take Isabel to move 10 m

a if Isabel and Gabriel weigh the same amount

b if Isabel weighs twice as much as Gabriel?

11 A ball of mass 100 g is projected along a straight line with a velocity of $3\,\mathrm{m\,s^{-1}}$. After 5 metres it meets another ball of equal mass. Given that it takes 2 seconds to return to its initial position, find the velocity of both balls after they collide, stating the direction of motion clearly.

3.3 Coalescing bodies

This diagram shows a train travelling at a constant speed on a horizontal track.

The total momentum of the train is the sum of the momentum of the carriages together with the momentum of the engine. If the carriages are decoupled while the train is travelling at constant speed, then the total momentum is still the same – the train does not lose momentum.

> Carriages carry people and trucks carry goods, but the mechanics is the same either way.

When objects travel at the same speed alongside each other you can combine their momenta. When two bodies collide and then come together to form one mass we say that they have **coalesced**.

Consider another situation.

A daughter, of mass 30 kg, is ice skating up to her mother, of mass 60 kg. Initially the daughter is moving at $6\,\mathrm{m\,s^{-1}}$ and the mother is moving in the same direction at $3\,\mathrm{m\,s^{-1}}$. Given that momentum is conserved, what happens after they collide if the daughter is moving at $4\,\mathrm{m\,s^{-1}}$?

$$m_A u_A + m_B u_B = m_A v_A + m_B v_B$$

Let the final velocity of the mother $= v_B$

So $30 \times 6 + 60 \times 3 = 30 \times 4 + 60 \times v_B$

$$v_B = \frac{(30 \times 6 + 60 \times 3 - 30 \times 4)}{60} = 4\,\mathrm{m\,s^{-1}}$$

This is the same velocity as the daughter. This tells you that the mother and daughter must be travelling together. For example, they may have joined hands and are now sliding along together.

$$m_A u_A + m_B u_B = m_A v_A + m_B v_B$$

If the two bodies coalesced, we know that $v_A = v_B$ so we can write v for the final velocity of the combined body. This gives

$$m_A u_A + m_B u_B = m_A v_A + m_B v_B = m_A v + m_B v = (m_A + m_B)v$$

So, for coalescing bodies,

$$m_A u_A + m_B u_B = (m_A + m_B)v$$

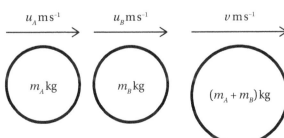

If two objects move at the same speed after colliding, then can you be certain that they have coalesced? If they hit head on, what would determine the final direction of the coalesced body?

Example 5

Two balls of clay, P and Q, are thrown towards each other such that they combine when they meet. If the mass of P is twice the mass of Q and, initially, the velocity is $9\,\text{m s}^{-1}$ for P and $-3\,\text{m s}^{-1}$ for Q, find the velocity of the combined ball.

Solution

Let the mass of $Q = m$ kg. This is a diagram of the situation.

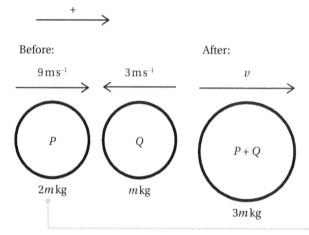

The mass of P is twice that of Q, so P has a mass of $2m$ kg.

Use the principle of conservation of momentum formula.

$$m_A u_A + m_B u_B = (m_A + m_B)v$$
$$2m \times 9 + m \times (-3) = 3m \times v$$
$$18m - 3m = 3mv$$
$$\text{so } 15m = 3mv$$
$$15 = 3v$$
$$v = 5\,\text{m s}^{-1}$$

Dividing by m simplifies the equation.

You are able to solve the problem without knowing the mass of any of the balls. What did you need to be told about their masses? In many situations it may be difficult or unnecessary to know the mass of an object, only how it compares to others. Is this also true of the velocity of the objects?

Example 6

On a rainy day, a car starts to skid. It crashes into the back of another car, and the two cars become joined together and then skid along the road. The police estimate that the car in front is larger, and has a mass approximately one and a half times that of the car at the rear.

The police question the drivers.

The driver of the car in front says that she was driving at $50\,\mathrm{km\,h^{-1}}$ before the crash and that directly after the crash she thinks that the two cars were sliding at $55\,\mathrm{km\,h^{-1}}$.

The driver of the rear car estimates that he was travelling at $45\,\mathrm{km\,h^{-1}}$ before the crash, and directly after the crash he thinks that the two cars were travelling at $50\,\mathrm{km\,h^{-1}}$.

For each situation, calculate the initial speed of the other car. Which driver's statement should the police believe?

Solution

Let the mass of the smaller car be $m\,\mathrm{kg}$ and the larger car be $1.5m\,\mathrm{kg}$.

From the statement of the driver in front (in the larger car),

$$v_r \times m + 50 \times 1.5m = 55 \times (2.5m)$$

> Using conservation of momentum and a combined mass of $m + 1.5m = 2.5m$.

Subtracting the initial momentum of the larger car and dividing by m gives

$$v_r = 55 \times 2.5 - 50 \times 1.5 = 62.5\,\mathrm{km\,h^{-1}}$$

> v_r is the final velocity of the car at the rear.

From the statement of driver in the rear (in the smaller car),

$$45 \times m + v_f \times 1.5m = 50 \times (2.5m)$$

> You could choose masses that are in the correct ratio, such as $1000\,\mathrm{kg}$ and $1500\,\mathrm{kg}$.

$$1.5v_f = 50 \times 2.5 - 45 = 80$$

So $v_f = \dfrac{80}{1.5} = 53\frac{1}{3}\,\mathrm{km\,h^{-1}}$.

> v_f is the final velocity of the car in front.

This second result would mean that the car at the rear, which skidded, was initially going more slowly than the car in front. If this was the case, then when the car skidded, it would not have caught up with the car in front in order to crash into it.

This means that the police should believe the statement of the driver of the car in front.

Mathematics in life and work: Group discussion

Manoeuvring in space can be very difficult for many reasons, such as a lack of gravity! However, the conservation of momentum is a very important principle that can help.

A capsule is docking with a spaceship. The capsule has a mass of 6000 kg, and is approaching the ship at a speed of $8001.5\,\mathrm{m\,s^{-1}}$. The space ship has a mass of 400 000 kg and is travelling at a speed of $8000\,\mathrm{m\,s^{-1}}$.

1   What is the momentum of each object, and the total momentum of the system?

2   Once the capsule meets the spaceship they link together. How has the speed of the spaceship changed? Make the direction of the ship clear in your answer.

3   It is usual to talk of the relative speed of an object. What is the speed of the capsule relative to the ship?

4   When the capsule approaches the spaceship, it tries to slow down to a relative speed of $0.5\,\mathrm{m\,s^{-1}}$ by use of a 'manoeuvring thruster', which blasts out 25 kg of fuel. At what (relative) speed does it need to do this?

Exercise 3.3A

1 These diagrams show two bodies which coalesce on impact. Calculate the missing mass or velocity in each case.

a

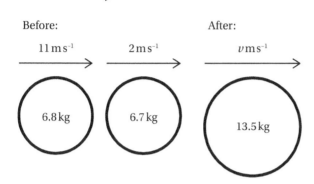

b

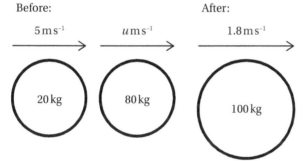

c

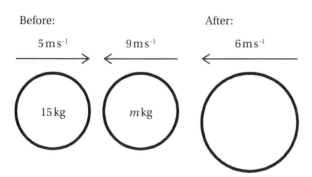

d

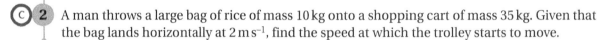

Before:

0.1 m s⁻¹ → u m s⁻¹ →   After: 0.031 m s⁻¹ →

120 g   1.38 kg

C **2** A man throws a large bag of rice of mass 10 kg onto a shopping cart of mass 35 kg. Given that the bag lands horizontally at 2 m s⁻¹, find the speed at which the trolley starts to move.

PS **3** An empty truck is coasting along a frictionless horizontal track. It starts to rain. State whether, as the rain collects in it, the truck will speed up, slow down or remain at the same speed, giving justification for your answer.

C **4** Lynda has a pet dog, Scruff, which has a mass of 2 kg. They are standing in the park when Scruff spots a cat. Scruff jumps up to chase after it. Just as his leash goes taut, Scruff is travelling at a velocity of 13 m s⁻¹. Find the common speed of Scruff and Lynda, given that Lynda has a mass of 63 kg.

MM **5** A bullet of mass 0.01 kg is fired into a block of wood of mass 2 kg at a speed of 540 m s⁻¹. The bullet is embedded into the block. Find the speed of the block immediately after the impact.

MM **6** A particle of mass $4m$ kg and velocity $2u$ m s⁻¹ strikes a particle of mass $2m$ kg. The two particles coalesce. What is the velocity of the combined particle, as a multiple of u, given that the $2m$ particle has the following initial conditions?

    a It is initially at rest.

    b Initially, the $2m$ particle is moving at u m s⁻¹.

    c What would happen if, initially, the $2m$ particle is moving at $3u$ m s⁻¹?

7 Given that two objects, A and B, with masses m_A and m_B and inital velocities u_A and u_B, coalesce and move with a common velocity, v, write an expression for m_A in terms of m_B, u_A, u_B and v.

PS **8** The diagram shows a ball of mass 1 kg moving in a tube with a velocity of 14 m s⁻¹ when a small internal explosion causes it to break into two pieces moving in opposite directions. One piece has mass 400 g and velocity 47 m s⁻¹ in the same direction as the ball was travelling. Find the immediate velocity of the second piece of the ball.

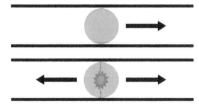

9 Sebastian makes a cart, which he can jump onto and move. He decides to test it on a smooth, flat road.

He stands on it, but it does not move. He tries again, jumping from a standing position and landing on the cart with a horizontal speed of 0.5 m s^{-1}. Sebastian has a mass of 20 kg, and the cart has a mass of 12 kg.

a Calculate the speed of the cart with Sebastian riding on it.

He tries a running start, starting 2 m away from the cart and accelerating from rest at a rate of 0.25 m s^{-2}.

b Calculate the new speed of the cart with Sebastian riding on it.

He wants the cart to go at a speed of 2 m s^{-1}.

c How far will he have to run at this same rate of acceleration, before jumping into the cart, to achieve this final speed? Is this a realistic situation?

10 Two balls collide as shown in the diagram.

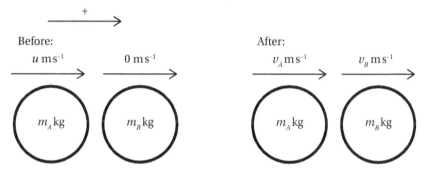

a Write an expression for v_B in terms of u, v_A, m_A and m_B.

A continues to travel towards B, and A has half the mass of B.

b Write an inequality for u in terms of v_B.

SUMMARY OF KEY POINTS

> The momentum of an object is the product of its mass and velocity. This is written $p = mv$. Momentum has units $kg\,m\,s^{-1}$

> If two objects directly collide, then they hit in a straight line, not at an angle, so it is a one-dimensional or linear situation.

> If two objects coalesce, then they join upon impact and move together.

> When there are no external forces on a system, the total momentum remains constant. This is called the principle of conservation of momentum.

> For two objects, A and B, that collide: $m_A u_A + m_B u_B = m_A v_A + m_B v_B$.

> For two objects, A and B, that coalesce: $m_A u_A + m_B u_B = (m_A + m_B)v$.

EXAM-STYLE QUESTIONS

1   Calculate the momentum of a ball of mass 0.5 kg and a velocity of $4.5\,m\,s^{-1}$. It collides with a block of mass 3 kg that is free to move. The block is initially stationary, but upon being hit it moves at $v\,m\,s^{-1}$. For which values of v does the ball rebound, changing direction?

2   An object has a momentum of $54\,kg\,m\,s^{-1}$.

   a   Given that it is travelling at $35\,km\,h^{-1}$ calculate the mass of the object.

   The object loses half of its mass, but its total momentum is conserved.

   b   Write down the new velocity of the object.

C 3   A car safety testing centre wishes to simulate the effect of a car of mass 1500 kg crashing head-on into a car of mass 1300 kg. Both cars are to be travelling at $36\,km\,h^{-1}$.

   a   What is the momentum of each car? Give your answer in $kg\,m\,s^{-1}$.

   The test centre only has the facilities to have one moving car, but can crash it into a wall. It chooses the 1500 kg car, and accelerates it to $v\,m\,s^{-1}$ just as it hits the wall.

   b   Given that it wishes to make the total magnitude of the momentum the same as if the two cars were meeting, calculate the value of $v\,m\,s^{-1}$.

MM 4   Particle A, of mass $2m$ and initial velocity $3u$, collides with particle B, of mass $3m$ and initial velocity $2u$.

   Calculate the final velocity of:

   a A if B has a final velocity of $4u$

   b A if B has a final velocity of $3u$

   c   both particles, given that they combine

   d B if A has a final velocity of $-3u$.

5 Three balls *A*, *B* and *C* of masses 1 kg, 2 kg and 3 kg are in a line on a smooth horizontal surface with initial velocities $1\,\mathrm{m\,s^{-1}}$, $0\,\mathrm{m\,s^{-1}}$ and $0\,\mathrm{m\,s^{-1}}$, as shown in the diagram, such that ball *A* is travelling towards ball *B*.

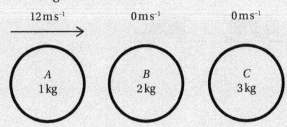

$12\,\mathrm{m\,s^{-1}}$ $0\,\mathrm{m\,s^{-1}}$ $0\,\mathrm{m\,s^{-1}}$

A 1 kg   *B* 2 kg   *C* 3 kg

a What is the initial momentum of ball *A*?

When *A* collides with *B*, ball *A* stops moving.

b What is the velocity of *B* when it is in motion?

Subsequently, when *B* collides with *C*, ball *B* stops moving.

c What is the final velocity of *C*?

d Write down the final momentum of ball *C*, explaining the significance of your answer.

6 Two identical beads are free to move along a smooth horizonal bar such that any collisions are direct.

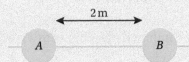

2 m

A   *B*

Initally, the beads are at rest 2 m apart. A force is applied to *A* for 3 seconds such that accelerates at a rate of $0.4\,\mathrm{m\,s^{-2}}$ towards bead *B*.

a Show that when the force is removed bead *A* has not reached bead *B*.

Bead *A* subsequently collides with bead *B*.

b Given that bead *A* stops moving after the collision, find the velocity of *B*.

7 a A basketball weighs 13 times more than a golf ball. If they are thrown at each other with an equal and opposite velocity, *u*, show that

$$v_\mathrm{g} = 12u - 13v_\mathrm{b}$$

where v_g is velocity of the golf ball after impact and v_b is the velocity of the basketball after impact.

b A bullet of mass *m* kg is fired from a gun of mass *M* kg, which is free to recoil. Let the initial velocities of the bullet and gun be $u\,\mathrm{m\,s^{-1}}$ and $U\,\mathrm{m\,s^{-1}}$, respectively. Show that the initial speeds of the bullet and gun are in inverse ratio to their masses.

8 An experiment is set up such that five trucks, each of mass 45 000 kg, are lined up along a smooth horizontal railway line, as shown in the diagram. Initially the trucks are all stationary. The first truck, A, is then speeded up to a velocity of $12\,\mathrm{m\,s^{-1}}$.

Truck A meets truck B and the two join together.

a Calculate the velocity of the two trucks.

The trucks continue to roll freely along the tracks, such that A and B meet C, then A, B and C meet D, and finally all five trucks travel together.

b Show that the final velocity of the five trucks is $2.4\,\mathrm{m\,s^{-1}}$.

9 Two objects A and B, of mass $4.5\,\mathrm{kg}$ and $m\,\mathrm{kg}$ approach each other at speeds of $4\,\mathrm{m\,s^{-1}}$ and $6\,\mathrm{m\,s^{-1}}$, respectively. They coalesce on impact.

a Calculate the value of m if the two objects cease to move after impact.

b Calculate the value of m if they move at $2.4\,\mathrm{m\,s^{-1}}$ in the direction of B after impact.

In fact, $m = 18\,\mathrm{kg}$.

c Calculate the final velocity of the combined mass, making the direction clear.

After a short time, the combined mass breaks apart into two sections with masses $10.5\,\mathrm{kg}$ and $12\,\mathrm{kg}$. The $10.5\,\mathrm{kg}$ section moves in the same direction as the combined mass but with a speed of $1\,\mathrm{m\,s^{-1}}$.

d Calculate the velocity of the second mass.

10 Two discs, X and Y, of mass m and $2m$, respectively, are sliding along a smooth horizontal surface. They collide directly. Prior to the collision, the velocity of X is $4u\,\mathrm{m\,s^{-1}}$ and the velocity of Y is $u\,\mathrm{m\,s^{-1}}$. Let the velocity of X after the collision be $v_x\,\mathrm{m\,s^{-1}}$ and the velocity of Y after the collision be $v_y\,\mathrm{m\,s^{-1}}$. $u > 0$.

a Show that $v_x = 6u - 2v_y$.

b Given that Y is now moving 150% faster than before the collision, work out the percentage decrease of the speed of X.

Before:

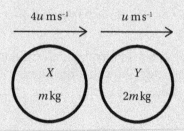

After:

$v_x\,\mathrm{m\,s^{-1}}$

$v_y\,\mathrm{m\,s^{-1}}$

11

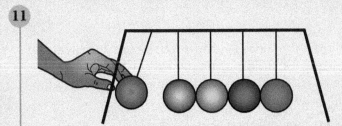

Noel makes a Newton's cradle which has five balls made from different materials, so that they are the same shape but different masses. They have masses in the ratio $5:4:3:2:1$ from left to right. He lifts up the heaviest one and it swings such that the velocity upon hitting the next-heaviest ball is 10 cm per second.

a Assuming that the heaviest ball then stops moving (as in a normal Newton's cradle), how fast will the lightest ball, at the opposite end, start to move?

Noel changes the position of the balls to the order $5:3:2:4:1$ and lifts the first ball in the same way.

b How fast does the lightest ball, at the opposite end, start to move in this situation? Justify your answer fully.

Finally, he changes the position of the balls to the order $3:5:1:2:4$ and lifts the first ball (now the middle-sized weight) in the same way.

c How fast does the last ball, at the opposite end, start to move in this situation?

12 A car of mass 2000 kg has broken down on a level crossing, 125 m from the end of a station platform, which is 160 m long. Everybody gets out the car, but the approaching train, from the other side of the station, cannot be contacted. It is a goods train of mass 20 000 kg which did not need to stop at the platform. Before reaching the platform, the train driver notices the car and applies the brake. A witness notices that it takes the front of the train 5 seconds to pass the platform and a further 5 seconds to reach the car. Calculate the momentum of the train as it hits the car and the speed of the combined car and train that assuming, they coalesce.

13 A bus of mass 10 tonnes is following a car of mass 1.2 tonnes at a distance of 30 m. Initially, both are travelling at a constant speed of 36 km h^{-1}. The driver's foot slips and causes the bus to accelerate uniformly. After 5 seconds, the bus crashes into the back of the car, and the two vehicles coalesce. Find the acceleration (in m s^{-1}) of the bus and the velocity (in km h^{-1}) of the vehicles immediately after the crash.

14 Ball A (of mass 2 kg) and ball B (of mass 3 kg) collide and coalesce. Their final velocity, after colliding, is 3 m s^{-1}. Ten seconds before they coalesced, both balls started from rest and they were accelerating at different rates. Given that ball B accelerated uniformly for the ten seconds at 0.1 m s^{-2}, find the uniform acceleration of ball A. How far apart were they ten seconds before they collided?

15 A railway engine of mass 25 000 kg is positioned 50 m behind two carriages, each of mass 15 000 kg, which are also 50 m apart. The resistances to forward motion are 2 kN on the engine and 0.5 kN on each of the carriages. The train is on a straight flat section of the track. The engine begins at rest, and starts to produce a driving force of 18 kN. As the engine meets the carriages, they instantaneously join together to form a larger train.

Calculate the final velocity of the engine after it has connected with the second carriage.

Mathematics in life and work

A space agency is designing a new way to increase the speed of deep-space probes, by repeatedly splitting the shuttle into smaller sections. Initially, a shuttle of mass 160 000 kg has a velocity of u km s^{-1}. The shuttle then splits into two sections of equal mass such that one of the sections is now moving at a velocity of u km s^{-1} in the opposite direction.

1 Show that the velocity of the second section of the shuttle is now $3u$ km s^{-1}.

This second piece then splits into two smaller sections of equal mass. Again the rear section has the same speed, but the opposite direction.

This continues until the mass of the shuttle is 20 000 kg.

2 Calculate the velocity of the shuttle at each step and hence find the final velocity of the shuttle of mass 20 000 kg as a multiple of u.

An alternative approach to speeding up a shuttle would be to eject the smallest shuttle, of mass 20 000 kg, directly from the largest shuttle, of mass 160 000 kg.

The final shuttle must attain the same speed by either method. $u = 140$ km s^{-1}.

3 Using this second approach, what is the velocity of the two sections after the first, and only, separation?

4 NEWTON'S LAWS OF MOTION

Mathematics in life and work

Isaac Newton (1642–1726) was a renowned physicist and mathematician who formulated the three laws of motion that bear his name. These laws demonstrate the relationship between the mass of an object and its acceleration. For example, the first law implies that without an external force acceleration will not happen and the second law states that where there is a resultant force it is proportional to both the mass and the acceleration. The third law was used extensively in **Chapter 1 Forces and equilibrium** and states that every action has an equal and opposite reaction. So, for example, when forces are resolved vertically for a stationary particle on a horizontal surface, the weight and reaction force balance.

There are many examples of careers where an understanding of Newton's laws and their implications is necessary – for example:

> If you were an architect drawing up plans for a lift in a hotel, you would need to consider the maximum load in the lift, the material used for the cables, the distance between floors and the velocity and acceleration of the lift whilst in motion.

> If you were designing a crane, you would need a sound understanding of pulleys and how they can be used to make lifting more efficient.

> If you were a car manufacturer, you would need to consider the weight of the materials used to make the car, the durability of the tyres and how they interact with the road, the strength of the driving force that could be exerted by the engine and the braking force that could be exerted by the brakes.

In this chapter, you will investigate Newton's laws of motion in the context of car manufacturing.

LEARNING OBJECTIVES

You will learn how to:

> apply Newton's laws of motion to the linear motion of a particle of constant mass moving under the action of constant forces, which may include friction, tension in an inextensible string and thrust in a connecting rod

> use the relationship between mass and weight

> solve simple problems that may be modelled as the motion of a particle moving vertically or on an inclined plane with constant acceleration

> solve simple problems which may be modelled as the motion of connected particles.

LANGUAGE OF MATHEMATICS

Key words and phrases you will meet in this chapter:

Newton's first law, Newton's second law, Newton's third law, pulley

PREREQUISITE KNOWLEDGE

You should already know how to:

> solve linear equations in one unknown and simultaneous linear equations in two unknowns

> construct and transform complicated formulae and equations

> identify the forces acting in a given situation

> use the principle that, when a particle is in equilibrium, the sum of the components in any direction is zero

> find and use the components and resultants of forces

> understand the concepts of limiting friction and limiting equilibrium, recall the definition of coefficient of friction, and use the relationship $F = \mu R$ or $F \leqslant \mu R$, as appropriate

> use appropriate formulae for motion with constant acceleration in a straight line.

You should be able to complete the following questions correctly:

1 Solve these simultaneous equations.

$$30 - y = 3x \qquad \text{①}$$
$$y - 10 = x \qquad \text{②}$$

2 Make x the subject of the equation $x \cos \theta - y \sin \theta = y \cos \theta + x \sin \theta$.

3 A 10 N box is pulled along a rough horizontal table at a steady speed by a string inclined at 26° to the horizontal. Given that the coefficient of friction between the box and the table is $\frac{1}{10}$, find the tension in the string.

4 A particle moving in a straight line with constant acceleration has an initial velocity of $4 \, \text{m s}^{-1}$ and travels 55 m in 5 s.

   a Find the acceleration of the particle.

   b Find the final velocity of the particle.

4.1 Newton's laws of motion

According to legend, in the late 17th century the mathematician Isaac Newton was sitting beneath an apple tree. After an apple fell on his head, he devised his universal law of gravitation. Since the initial velocity of the apple was zero, but the final velocity was not, the apple had accelerated to the ground. This indicated that a force must be acting on the apple – Newton named this force gravity. He then surmised that the same force of gravity that made the apple accelerate towards the Earth must affect other objects, such as the Moon. In turn, the Sun exerts a gravitational force upon the Earth. In fact, any two objects exert a gravitational force on each other.

Newton's first law states that a particle will remain at rest or continue to move with a constant velocity in a straight line unless acted upon by an external force.

In the case of the apple falling from the tree, the apple is motionless until it becomes detached from the tree, at which point it accelerates to the Earth.

Newton's second law states that the resultant force, F, acting upon a particle is proportional to the particle's mass, m, and acceleration, a. This results in the formula $F = ma$.

This extends the first law by stating that when an external force is applied, the particle will no longer be stationary or travelling at a steady speed, but instead will accelerate. The rate of acceleration is dependent on the magnitude of the force and the mass of the particle.

Newton's third law states that every action has an equal and opposite reaction.

If Newton's third law was not true, then the apple would not stop once it hit the ground. As observed in **Chapter 1 Forces and equilibrium**, you know that your weight is balanced by a force from the ground, which prevents you from falling into it and this is true for any object on a stable, solid surface, such as a vase on a table or a car on a road.

KEY INFORMATION

You should know and be able to apply Newton's three laws of motion.

Types of force

In this chapter, you will need to use the five types of force you met in **Chapter 1 Forces and equilibrium**: weight, normal reaction, tension, thrust and friction. Any other forces such as the driving force of an engine or air resistance will be indicated.

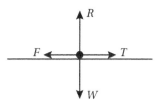

In this diagram, the reaction and weight forces are balanced and there is no vertical motion. If the particle is travelling at a steady speed along the horizontal line so that it is in equilibrium (see **Chapter 1 Forces and equilibrium**), then the tension and friction are also balanced. This is confirmed by Newton's first law, where the particle remains at a steady speed unless an external force is applied (the resultant force is zero). If the tension is greater than the friction, then Newton's second law ($F = ma$) states that the resultant force ($T - F$) will equal ma and the particle will accelerate.

As well as the new content in this chapter, you can expect questions to test your knowledge of forces from **Chapter 1 Forces and equilibrium** and the equations of uniformly accelerated motion from **Chapter 2 Kinematics of motion in a straight line**, as you will see in **Example 2**.

Mass and weight

Based upon the formula $F = ma$, weight is given by the formula $W = mg$, where m is mass in kg and g is the acceleration due to gravitational attraction, which on Earth is approximately $10\,\mathrm{m\,s^{-2}}$, as you used in **Chapter 2 Kinematics of motion in a straight line**. This value of g is only appropriate to objects on the surface of the Earth and the value will vary according to the object and the distance from the centre of the Earth. For example, 1000 km above the Earth's surface, the value of g is $7.33\,\mathrm{m\,s^{-2}}$. On the surface of Mars, g is $3.75\,\mathrm{m\,s^{-2}}$ and on the Moon it is $1.6\,\mathrm{m\,s^{-2}}$. On Jupiter, due to its huge mass, g is $26.0\,\mathrm{m\,s^{-2}}$ – almost three times greater than on Earth!

Modelling assumptions

Throughout this chapter, you will use simplified models to represent real-life situations. Each section includes a brief explanation of the modelling assumptions you will be using. In this first section, for example:

> Objects will be modelled as particles so that their mass is concentrated at a single point.

> Motion will take place horizontally in a straight line (such as along a straight horizontal road) so that angles need not be considered and so that weight has no effect in the direction of motion, since it will be perpendicular to the motion.

Note that weight is different from mass because it depends on the value of gravitational acceleration, g.

KEY INFORMATION

$W = mg$

KEY INFORMATION

All questions in this book take g as $10\,\mathrm{m\,s^{-2}}$.

Example 1

A boy is pulling his sledge horizontally across the snow by means of a light rope which is also horizontal. The frictional force is 40 N and the sledge is accelerating at $0.3\,\mathrm{m\,s^{-2}}$. The sledge has a mass of 18 kg. Find the tension in the rope.

Solution

Draw a diagram. Model the sledge as a particle. There are vertical weight and reaction forces and horizontal tension and friction forces. Since the mass is given rather than the weight, use the formula $W = mg$ to write the weight as $18g$ N.

Because the sledge is accelerating, the tension and friction are not balanced and instead there is a resultant force F equal to ma. Represent the acceleration $a\,\mathrm{m\,s^{-2}}$ by a double-headed arrow pointing right.

When you use the formula $F = ma$, where F is the resultant force, you should add up the forces in the direction of motion. In this example, the sledge is moving to the right.

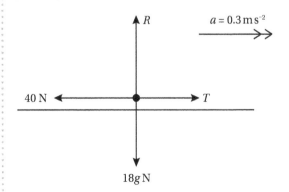

Consider the horizontal forces. Since the sledge is moving to the right, the tension is positive and the resistance is negative.

Substituting into $F = ma$,

$T - 40 = 18 \times 0.3$

$ = 5.4$

$ T = 45.4\,\text{N}$

Example 2

A car is towing a caravan along a straight horizontal road at $24\,\text{m s}^{-1}$. The car decelerates at a constant rate for $112\,\text{m}$, which takes $8\,\text{s}$. Given that the caravan has a mass of $400\,\text{kg}$ and that the coefficient of friction between the caravan and the road is $\frac{1}{8}$, find the thrust in the tow bar whilst the car is decelerating.

Solution

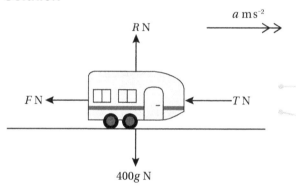

It is a good idea to draw a diagram to represent the information.

Remember that the thrust force acts *towards* the object.

You need to use the formula $F = ma$. The resultant force F will be a combination of the resistance to motion and the thrust in the tow bar. The mass m is $400\,\text{kg}$. The acceleration a can be found from the equations of uniformly accelerated motion.

Note that F is used for both the resultant force and the friction.

You are given that $u = 24\,\text{m s}^{-1}$, $s = 112\,\text{m}$ and $t = 8\,\text{s}$.

The relevant equation of uniformly accelerated motion is $s = ut + \frac{1}{2}at^2$.

Substitute into the equation.

$112 = 24 \times 8 + \frac{1}{2} \times a \times 8^2$

$ = 192 + 32a$

$-80 = 32a$

$ a = -2.5\,\text{m s}^{-2}$

Resolve horizontally (in the direction of motion).

$R(\rightarrow)$

$-T - F = ma$

Substitute into $F = ma$.

$-T - F = 400 \times -2.5 = -1000$

Resolve vertically.

$R(\uparrow)$

$R - 400g = 0$

Substitute $F = \mu R$.

$-T - \frac{1}{8} \times 400g = -1000$

$-T - 50g = -1000$

$-T = -1000 + 50g = -1000 + 50 \times 10$

$\quad = -500\,N$

The thrust is 500 N.

> When you use the formula $F = ma$, where F is the resultant force, you should add up the forces in the direction of motion. In this example, the caravan is moving to the right, but both horizontal forces are pointing to the left, so both are negative.

Exercise 4.1A

1 **a** Find the magnitude of the resultant force acting on a particle of mass 800 g that would produce an acceleration of $3.1\,m\,s^{-2}$.

   b A particle of mass m kg has a deceleration of $4\,m\,s^{-2}$ when acted upon by a resultant force of 5.2 kN. Find the value of m.

2 **a** Find the weight of a dog of mass 8.4 kg on the Earth's surface.

   b An astronaut weighs 735 N on the surface of Earth. Find the weight of the astronaut on the surface of Mars ($g = 3.75\,m\,s^{-2}$).

3 A car travelling along a straight horizontal road produces a driving force of 1.27 kN whilst experiencing a frictional force of 190 N. The car takes 15 s to accelerate uniformly from $4\,m\,s^{-1}$ to $22\,m\,s^{-1}$. Find the mass of the car.

4 A particle at rest 2 m from the edge of a horizontal table is subjected to a thrust of X N.

   The coefficient of friction is $\frac{1}{3}$. The mass of the particle is 75 kg.

   Find the value of X given that:

   a the particle does not accelerate

   b the particle accelerates at $2\,m\,s^{-2}$

   c the particle falls off the edge of the table after 0.5 seconds.

5 A minibus of mass 1500 kg produces a driving force of 1125 N. Given that the initial velocity of the minibus is $4\,m\,s^{-1}$, find the time it takes the minibus to travel 385 m along a straight horizontal road. The coefficient of friction between the minibus and the road is 0.035.

6 A car of mass 960 kg accelerates from $15\,\mathrm{m\,s^{-1}}$ to $30\,\mathrm{m\,s^{-1}}$ along a straight 400 m stretch of a horizontal motorway at a constant rate. During this acceleration, the engine produces a driving force of 1.24 kN. The motorist then spots a fault on the dashboard and decides to pull over onto the hard shoulder, removing the driving force. The car then decelerates at a constant rate with a braking force of 500 N. Assuming that the frictional force is constant throughout, how much further does the car travel before it comes to rest?

7 Starting from rest, a 1.6 tonne coach travels along a straight horizontal lane. The engine exerts a driving force of 5.9 kN. The coach experiences a uniform frictional force of 1.1 kN.

After 11 s, the driver spots a roadblock 260 m ahead. He immediately decelerates at a constant rate, reducing the driving force to zero and applying a braking force, and comes to rest 20 m before the roadblock.

Find the braking force required to bring the coach to a halt.

 8 A remote-controlled truck moves along a rough horizontal floor. The coefficient of friction between the floor and the truck is $\frac{3}{8}$ and the driving force is 54 N. When the truck is placed onto a different horizontal surface, for which the coefficient of friction between the surface and the truck is $\frac{2}{3}$, it requires a driving force of 89 N to maintain the same acceleration.

a Find the mass of the truck.

b Find the acceleration.

 9 A trolley is pulled across a rough horizontal car park by a rope inclined at an angle α to the horizontal. When $\cos\alpha = \frac{20}{29}$, the coefficient of friction between the trolley and the car park is $\frac{17}{40}$ and the trolley accelerates at $0.2\,\mathrm{m\,s^{-2}}$. When instead $\sin\alpha = \frac{20}{29}$, the normal reaction force between the trolley and the car park is 180 N.

Assuming that the tension is constant:

a find the tension in the rope

b find the mass of the trolley.

 10 A crate of mass 20 kg is pulled across a rough horizontal yard by means of a light handle inclined at angle θ to the horizontal, where $\sin\theta = \frac{3}{5}$. The coefficient of friction between the crate and the surface of the yard is 0.25. The tension in the handle is 60 N. The initial velocity of the crate is $4\,\mathrm{m\,s^{-1}}$. The total time taken to pull the crate 1.44 km is given by t s.

a Show that $7t^2 + 160t = 57\,600$.

b Hence find the value of t.

4.2 Vertical motion

In **Section 4.1**, where cars and particles were assumed to be travelling horizontally, you did not need to consider the weight of the object as a force because it acted perpendicularly (vertically) to the motion. However, when an object is travelling vertically then you do need to consider its weight. When an object travels upwards, the weight acts in the opposite direction and slows the object down, and when an object travels downwards, the weight is acting in the same direction and speeds the object up.

Modelling assumptions

❯ Motion is vertical, so angles need not be considered and weight contributes fully to the resultant force.

❯ The value of g can be considered to be constant over small changes in altitude (height).

Example 3

A suitcase of mass 4 kg is in freefall. It experiences air resistance of 0.2 N as it falls. Find the acceleration of the suitcase.

Solution

Start by drawing a diagram. Both the air resistance and the weight of the suitcase act vertically.

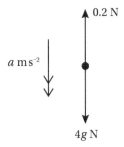

Consider the resultant force in the direction of motion (downwards).

$$F = ma$$
$$4g - 0.2 = 4a$$
$$4 \times 10 - 0.2 = 4a$$
$$39.8 = 4a$$
$$a = 9.95 \, \text{m s}^{-2}.$$

In most situations, you should assume that air resistance is negligible and hence acceleration in freefall is the same as g.

Example 4

A ball of mass 0.7 kg is raised vertically by a rope. Initially, the ball is stationary and at a height of 0.5 m above the ground. The tension in the rope is 10 N. After 1.5 s, the rope snaps. Find how long it takes, after the rope snaps, for the ball to return to the ground.

Solution

Draw a diagram.

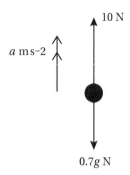

The motion of the ball goes through two stages, before and after the rope snaps. Before the rope snaps, the acceleration of the ball is determined by the tension in the rope and the weight of the ball. After the rope snaps, the acceleration is determined only by the weight of the ball, since the ball is in freefall.

During the first stage, the ball is acted upon by the upwards tension of 10 N and its weight, $0.7g$ N.

Hence the resultant force, F, is given by $(10 - 0.7g)$ N.

Substitute $F = (10 - 0.7g)$ N and $m = 0.7$ kg into the equation $F = ma$.

$(10 - 0.7g) = 0.7a$

$$a = \frac{10 - 0.7g}{0.7} = 4.29\,\text{m s}^{-2}$$

In order to find the time during the second stage once the rope has snapped, it is necessary to know the initial velocity of the ball, which will be the same as the final velocity when the rope snaps, and the height of the ball.

$u = 0\,\text{m s}^{-1}$, $t = 1.5\,\text{s}$, $a = 4.29\,\text{m s}^{-2}$

$v = u + at$

$\quad = 0 + 4.29 \times 1.5$

$\quad = 6.44\,\text{m s}^{-1}$

$\quad = ut + \frac{1}{2}at^2$

$\quad = 0 \times 1.5 + \frac{1}{2} \times 4.29 \times 1.5^2$

$s = 4.83\,\text{m}$

The height above the ground $= 4.83 + 0.5 = 5.33\,\text{m}$.

When the rope snaps, taking upwards as the positive direction, $u = 6.44\,\text{m s}^{-1}$, $s = -5.33\,\text{m}$, $a = -10\,\text{m s}^{-2}$.

$v^2 = u^2 + 2as$

$\quad = 6.44^2 + 2 \times -10 \times -5.33$

$\quad = 148$

$v = -12.2\,\text{m s}^{-1}$

Substitute into $v = u + at$ to find t.

$-12.2 = 6.44 - 10t$

$\quad t = 1.86\,\text{s}$

There are two answers for the equation $v^2 = 148$, which are $v = \pm 12.2$. However, since the suitcase is in freefall, it is travelling downwards, so v is negative because upwards has been chosen as the upwards direction.

Exercise 4.2A

1 A ball of mass 120 g is dropped from the top of a tower of height 25 m.

The ball experiences an air resistance of 0.03 N as it falls.

Find how much longer it takes for the ball to hit the ground than if air resistance were ignored.

2 A toy of mass 350 g is dropped in a swimming pool and hits the surface of the water before falling vertically to the bottom of the pool. The depth of the pool is 1.9 m. The toy experiences a resistance to its motion of 2.8 N. The initial velocity of the toy as it hits the water is 3.7 m s^{-1}. Find the time it takes the toy to reach the bottom of the pool.

3 A box of mass M kg is held at the top of a vertical rod. When the box is lowered vertically with an acceleration of 0.4 m s^{-2}, the compression in the rod is 12 N. Find the compression in the rod when the box is raised vertically with an acceleration of 0.4 m s^{-2}.

4 A 20 kg rock is pulled vertically upwards by a vertical cable from a speed of 2 m s^{-1} to a speed of 5 m s^{-1}. The tension in the cable is 250 N. Calculate how far the rock is pulled.

5 A 3 N weight at rest on the floor is attached to a vertical string, which pulls it vertically upwards. When the weight is 2.5 m above the floor, it is moving at 4 m s^{-1}.

a Find the tension in the string.

The string is then released.

b Find the maximum height achieved by the weight.

c Find the velocity of the weight when it returns to the floor.

6 A 28 kg block is projected vertically upwards from the ground at 7 m s^{-1}. When the block reaches a height of 2 m, it experiences a force of 420 N vertically upwards until it is 18 m above the ground. The force is then removed. Assume that air resistance can be ignored.

a Find the total time taken for the block to reach the height of 18 m.

b Find the speed of the block when it returns to the ground.

7 A diver of mass 50 kg drops from the 10 m board at the swimming pool. As the diver drops, she experiences air resistance of magnitude 10 N. When she goes under water, the water resistance is of magnitude 2500 N. Let g be 9.8 m s^{-2}.

a Find the maximum distance the diver travels under the surface of the water.

A second diver, of mass 60 kg, copies the first diver. Assume that the resistances are the same.

b How much further does the second diver travel under the surface of the water?

8 Two rocks fall from a cliff, 45 m above a pool of mud. When the rocks are in the mud, they experience a resistive force of 3.25 kN. One rock is 6 kg heavier than the other and the final position of this rock is 25 cm lower than for the other. Assume that air resistance can be ignored.

Find the mass of the lighter rock.

Why does the tension in a cord have a different magnitude depending on whether the load is moving upwards or downwards?

4.3 Inclined planes

When a particle is on an inclined plane, problems can require the use of the formulae $F = ma$ and $F = \mu R$, the equations of uniformly accelerated motion and resolving with angles. Sometimes a force will be removed or the angle will be increased. The value of μ will not change but $F = \mu R$ will still be valid for new values of F and R.

Example 5

A horizontal force of 2 N is just sufficient to prevent a block of mass 1 kg from sliding down a rough plane inclined at an angle of $\sin^{-1} \frac{7}{25}$ to the horizontal.

a Find the value of the coefficient of friction.

The horizontal force is removed.

b Find the acceleration with which the block will now move.

Solution

a Draw a diagram to represent this information.

This force diagram has four forces: the weight of g N, the reaction force, the frictional force and the horizontal 2 N force. Since the block is on the point of sliding down the plane, the friction acts up the plane. Let the angle of inclination be θ.

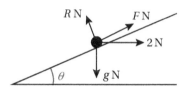

The phrase 'just sufficient to prevent' indicates that the block is in limiting equilibrium and hence $F = \mu R$.

Resolve parallel to the slope to find an expression for F.

R($\nearrow$)

$F + 2\cos\theta - g\sin\theta = 0$

Therefore $F = (g\sin\theta - 2\cos\theta)$ N.

Resolve perpendicular to the slope to find an expression for R.

R($\nwarrow$)

$R - g\cos\theta - 2\sin\theta = 0$

Therefore $R = (g\cos\theta + 2\sin\theta)$ N.

$\mu = \dfrac{F}{R} = \dfrac{g\sin\theta - 2\cos\theta}{g\cos\theta + 2\sin\theta}$

Look out for different ways of expressing the same information for limiting equilibrium such as 'just sufficient to prevent', 'on the point of slipping' and 'about to move'.

θ is given as $\sin^{-1}\frac{7}{25}$. Hence $\sin\theta = \frac{7}{25}$ and you need to find an expression for $\cos\theta$. Sketch a right-angled triangle with an opposite of 7 and a hypotenuse of 25.

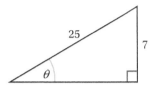

Hence $\sqrt{25^2 - 7^2} = \sqrt{576} = 24$ and

$\cos\theta = \dfrac{\text{adjacent (adj)}}{\text{hypotenuse (hyp)}} = \dfrac{A}{H} = \dfrac{24}{25}$.

From which

$$\mu = \frac{g\left(\frac{7}{25}\right) - 2\left(\frac{24}{25}\right)}{g\left(\frac{24}{25}\right) + 2\left(\frac{7}{25}\right)} = \frac{7g - 48}{24g + 14} = 0.0866.$$

b Draw a new diagram without the horizontal force to represent the change in information. Add in a double-headed arrow to indicate that the block will now accelerate down the plane.

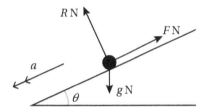

> If the original situation is adapted, then it is a good idea to draw a new force diagram.

Resolve down the slope in the direction in which the block will move.

$R(\swarrow)$

$g\sin\theta - F = ma = a$

> Note that since the block has a mass of 1 kg, ma can be rewritten as a.

In order to find the acceleration, it is necessary to find an expression for F.

Since $F = \mu R$, resolve perpendicular to the plane to find a new expression for R.

> Note that since the 2 N force has been removed, the reaction force, R, has also changed.

$R(\nwarrow)$

$R - g\cos\theta = 0$

Therefore

$R = g\cos\theta = \frac{24}{25}g\,\text{N}$.

Using the value of μ obtained in **part a**, 0.0866,

$F = 0.0866 \times \frac{24}{25}g = 0.831\,\text{N}$.

Hence

$a = g\sin\theta - F = \frac{7}{25}g - 0.831 = 1.97\,\text{m s}^{-2}$.

Example 6

A particle of mass 5 kg is just prevented from slipping down a slope by a 3 N force acting parallel to the slope. The slope is inclined at 20° to the horizontal.

Find:

a the coefficient of friction

b the acceleration when the 3 N force is removed

c how long it takes the particle to roll 10 m.

Solution

a Draw the force diagram.

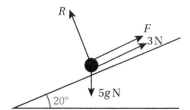

Resolve parallel to the slope.

R($\nearrow$)

$F + 3 - 5g\sin 20° = 0$

So $F = (5g\sin 20° - 3)$ N

Resolve perpendicular to the slope.

R($\nwarrow$)

$R - 5g\cos 20° = 0$

Therefore $R = 5g\cos 20°$ N.

$$\mu = \frac{F}{R} = \frac{5g\sin 20° - 3}{5g\cos 20°} = 0.300$$

b Draw an updated force diagram.

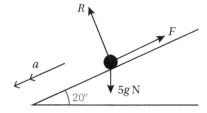

Resolve parallel to the slope in the direction of motion.

R($\swarrow$)

$5g\sin 20° - F = 5a$

The reaction force R will not have changed since the 3 N force that was removed was parallel to the slope.

Hence $R = 5g\cos 20°$ and $F = \mu R = 0.300 \times 5g \cos 20°$.

$5a = 5g \sin 20° - 0.300 \times 5g \cos 20° = 3.01$

$a = \dfrac{3.01}{5} = 0.602 \, \text{m s}^{-2}$

c   Since $a = 0.602 \, \text{m s}^{-2}$, $u = 0 \, \text{m s}^{-1}$ and $s = 10 \, \text{m}$, an equation of uniformly accelerated motion can be used to find the length of time it takes the particle to roll 10 m.

Substitute into $s = ut + \frac{1}{2}at^2$.

$10 = 0 + \frac{1}{2} \times 0.602 t^2$

$t^2 = 33.22$

$t = 5.76 \, \text{s}$

Exercise 4.3A

1   A tile of mass 900 g slides down a sloped roof. The roof is inclined at 26° to the horizontal.

Given that the coefficient of friction between any tile and the roof is 0.35,

a   find the acceleration of the tile.

A second tile of mass 800 g slides down the roof.

b   State the acceleration of the second tile.

2   A basket of mass 4 kg is pushed up a smooth plane by a force of 26 N, along the line of greatest slope. The plane is inclined at an angle θ to the horizontal. If the basket is accelerating at $0.5 \, \text{m s}^{-2}$, find the value of $\sin \theta$.

3   A ball of mass 3 kg is released from rest on a rough surface inclined at $\sin^{-1} 0.6$ to the horizontal. After 2.5 s, the ball has acquired a velocity of $4.9 \, \text{m s}^{-1}$ down the surface. Find the coefficient of friction between the body and the surface.

4   A body of mass 5 kg is initially at rest at the bottom of a rough inclined plane of length 6.3 metres. The plane is inclined at 30° to the horizontal.

The coefficient of friction between the body and the plane is $\dfrac{1}{2\sqrt{3}}$.

A constant horizontal force of magnitude $35\sqrt{3}$ N is applied to the body causing it to accelerate up the plane.

Find:

a   the time taken for the body to reach the top of the plane

b   the speed of the body on arrival at the top of the plane.

5   A box of mass 5 kg is pulled along rough horizontal ground by means of a rope inclined at $\sin^{-1} \dfrac{44}{125}$ to the ground. When the tension in the rope is 10 N the box is moving at constant speed.

a   Calculate the coefficient of friction between the box and the ground.

b   Calculate the acceleration of the box if the tension is increased to 15 N.

6 A particle of mass 2 kg rests in limiting equilibrium on a plane inclined at 25° to the horizontal. The angle of inclination is decreased to 20° and a force of magnitude 20 N is applied up a line of greatest slope.

    a Find the particle's acceleration.

When the particle has been moving for 2 seconds the force is removed.

    b Determine the furthest distance the particle will move up the plane.

7 A body of mass 500 g is placed on a rough plane which is inclined at 40° to the horizontal. The coefficient of friction between the body and the plane is 0.6.

    a Find the maximum value of the frictional force.

    b Show that motion will occur, stating the acceleration of the body.

8 A horizontal force of 4 N is just sufficient to prevent a block of mass 2 kg from sliding down a rough plane inclined at 23° to the horizontal. If the 4 N force is replaced by a force of X N, then the block is on the point of moving up the plane.

    a Find the value of X.

    b Find the acceleration if the X N force acts parallel to the slope rather than horizontally.

9 A particle of mass 2 kg slides down two rough planes. The first plane is inclined at α to the horizontal, where $\sin \alpha = \frac{3}{5}$ and the second plane is inclined at β to the horizontal, where $\sin \beta = \frac{7}{25}$. The coefficient of friction, μ between the particle and each plane is the same. Given that the acceleration of the particle on the first plane is ten times greater than that on the second, find the value of μ.

10 A block is projected up a rough plane at 6 m s^{-1}, coming to instantaneous rest after 8 m. The plane is inclined at 7.5° to the horizontal.

    a Find the coefficient of friction between the block and the plane.

    b Find the total time taken for the block to return to its original position.

Mathematics in life and work: Group discussion

You work for a car manufacturer. You have been asked to minimise the driving force that needs to be produced by the engine of a car when it is travelling uphill.

1 Why might it be a good idea to minimise the driving force?

2 Why are these factors important? How is the driving force affected by the mass of the car? How is the driving force affected by the coefficient of friction? How might you seek to minimise or maximise each one? What would you need to change?

3 What other factors would you need to consider to minimise the driving force? What could you increase or decrease in each case?

4 For a hill inclined at an angle of tan^{-1} 0.1, investigate the driving force required for masses of car between 1000 kg and 3000 kg and coefficients of friction between 0.03 and 0.07. Do your findings support your answers to **question 2** about how the driving force is affected by the mass or the coefficient of friction? If not, why not? What happens if you increase or decrease the other factors you suggested for **question 3**?

4.4 Connected particles

A car is travelling along a straight horizontal road. If the car is accelerating, then the engine exerts a driving force, D N. If the road is rough, then there will be a frictional force in the opposite direction. If the road is smooth, there will be *no* frictional force, but there may be other resistive forces to consider, such as air resistance. Assume that these resistive forces can be grouped as one overall resistance to motion, X N. The resultant force will be the difference between the driving force and this resistance to motion.

This is Newton's second law.

$F = ma$

$D - X = ma$

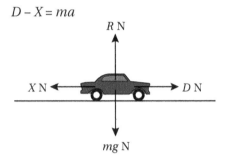

The car will also have a vertical weight force, $W = mg$ N, acting vertically downwards and a reaction force, R N, exerted on the car by the road. These forces will be equal and opposite.

This is Newton's third law.

Now consider what would happen if the car was towing a caravan. As well as the driving force and any resistive forces opposing the motion, there will be a tension force in the tow bar.

$F = ma$

$D - X_C - T = m_C a$ ①

Taken individually, without knowing the magnitude of the tension force, it may not be possible to calculate the other forces or the acceleration.

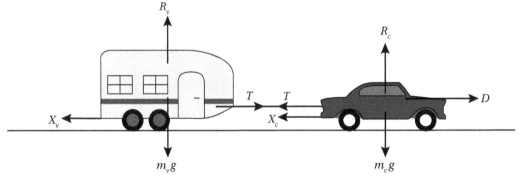

However, when you consider the horizontal forces acting on the caravan, there will be an equal and opposite tension force as well as the caravan's own resistive forces.

$F = ma$

$T - X_V = m_V a$ ②

Hence you have two equations, ①for the car and ②for the caravan, which can be solved simultaneously. If you know the masses of the vehicles and the driving and resistive forces, then the two unknowns are T and a. Note that, because the tensions are equal and opposite, adding these equations together will eliminate T.

$D - X_C - T = m_C a$ ①

$T - X_V = m_V a$ ②

Add ① + ②:

$D - X_C - T + T - X_V = m_C a + m_V a$

$\qquad D - X_C - X_V = m_C a + m_V a$

$\qquad\quad D - (X_C - X_V) = (m_C + m_V)a$

> **KEY INFORMATION**
> Simultaneous equations will usually be required to solve problems on connected particles.

> This equation is equivalent to applying Newton's law to the car and caravan together.

Modelling assumptions

> ⟩ Again, motion is strictly horizontal or vertical.

> ⟩ Tensions are equal and opposite.

> ⟩ Cables and tow bars are modelled as light inextensible strings or rods:

>> ⟩ light so that the tension is the same for both parts of the string

>> ⟩ inextensible so that the acceleration is the same for both particles.

Example 7

A car of mass 700 kg tows a trailer of mass 200 kg along a straight horizontal road. The engine of the car exerts a driving force of 1220 N. The car experiences a resistance to its motion of 200 N whilst the trailer experiences a resistance of 300 N. Find the acceleration of each vehicle and the tension in the tow bar.

Solution

As usual, it is best to represent the information in a diagram before you start.

The car is towing the trailer so in the diagram the vehicles are accelerating to the right.

Let the tension be T N, the acceleration be a m s^{-2} and the reaction forces be R_1 N and R_2 N. Note that, although they are not required in this question, they are equal and opposite to the weights of each vehicle.

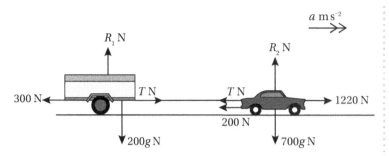

The resultant force on the car is given by
$(1220 - T - 200) = (1020 - T) \, \text{N}$.

Substitute into $F = ma$.

$1020 - T = 700a$ ①

The resultant force on the trailer is given by $(T - 300) \, \text{N}$.

Substitute into $F = ma$.

$T - 300 = 200a$ ②

Note that ① and ② are simultaneous equations and adding them together will eliminate T.

$(1020 - T) + (T - 300) = 700a + 200a$

$1020 - T + T - 300 = 900a$

$720 = 900a$

$a = 0.8 \, \text{m s}^{-2}$

You can now substitute $a = 0.8 \, \text{m s}^{-2}$ into equation ② to find the value of T.

$T - 300 = 200 \times 0.8$

$T - 300 = 160$

$T = 460 \, \text{N}$

Since each vehicle is accelerating at the same rate, the acceleration of each vehicle is $0.8 \, \text{m s}^{-2}$ and the tension in the tow bar is $460 \, \text{N}$.

Example 8

Two bricks, A and B, of masses $3 \, \text{kg}$ and $2 \, \text{kg}$, respectively are connected by a light cable. Initially the bricks are held such that the cable is vertical, with B $1 \, \text{m}$ above the ground. A vertical force of $67 \, \text{N}$ is applied to A, which causes the bricks to accelerate upwards. When B is $3 \, \text{m}$ above the ground the cable breaks. Show that B will travel a further $0.68 \, \text{m}$ before coming to instantaneous rest.

Solution

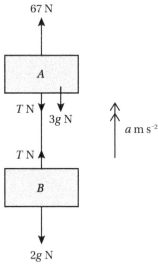

In this example, the motion is vertical, so the weights will contribute to the resultant forces.

The resultant force on A is given by $(67 - T - 3g)$ N.

Substitute into $F = ma$.

$67 - T - 3g = 3a$ ①

The resultant force on B is given by $(T - 2g)$ N.

Substitute into $F = m$.

$T - 2g = 2a$ ①

Again, as in the previous example, these are simultaneous equations and adding them eliminates T.

$(67 - T - 3g) + (T - 2g) = 3a + 2a$

$67 - 3g - 2g = 5a$

$17 = 5a$

$a = 3.4\,\mathrm{m\,s^{-2}}$

The bricks move 2 m before the cable breaks.

> When the cable breaks, the tension force is removed.

$u = 0\,\mathrm{m\,s^{-1}}, s = 2\,\mathrm{m}, a = 3.4\,\mathrm{m\,s^{-2}}$

$v^2 = u^2 + 2as$

$\quad = 0^2 + 2 \times 3.4 \times 2$

$\quad = 13.6$

$v = \sqrt{13.6}\,\mathrm{m\,s^{-1}}$

Once the cable breaks, B will continue to move upwards but will slow down because the gravitational acceleration acts downwards.

$u = \sqrt{13.6}\ \text{m s}^{-1},\ v = 0\ \text{m s}^{-1},\ a = -10\ \text{m s}^{-2}$

$v^2 = u^2 + 2as$

$s = \dfrac{v^2 - u^2}{2a}$

$\quad = \dfrac{0^2 - \left(\sqrt{13.6}\right)^2}{2 \times -10}$

$\quad = \dfrac{-13.6}{-20}$

$\quad = 0.68\ \text{m}$

Consider a car towing a trailer along a rough horizontal road. A typical force diagram would look like this.

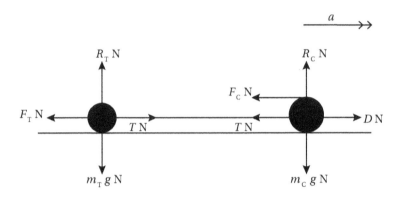

The car is towing the trailer to the right. The engine of the car is exerting a driving force. The tension in the tow bar is represented in the usual way with equal and opposite tensions from each of the vehicles. Each vehicle has its own weight, friction and reaction forces labelled with subscripts (C for car and T for trailer) and the friction and reaction are related by $F = \mu R$.

Note that if the vehicles become uncoupled, they act as if they are no longer connected. In this situation, there are two stages to the motion.

The first stage is when the vehicles are coupled and the particles are accelerating together.

The second stage is when the vehicles are uncoupled and the one that is being pulled is slowed down by friction and/or gravity.

It is important to understand how the two stages differ. When the vehicles are uncoupled, the tension force is removed and, as a result, the vehicle that is producing the driving force will find itself accelerating at a higher rate – since the resultant force

KEY INFORMATION

If the connection between two particles is impeded, such as vehicles becoming uncoupled or a string going slack or breaking, the particles act as if they are no longer connected.

is greater and the vehicle that is being pulled will find itself decelerating as a result of friction and/or its weight. In real life, there will always be a frictional force, however smooth a surface might appear. It is also worth noting that the final velocity for both vehicles from the first stage will be the initial velocity for both vehicles for the second stage.

Example 9

Two particles, A and B, of masses 2 kg and 3 kg, respectively, are connected by a light inextensible 7 m string. A 15 N force is applied to particle B, which accelerates the particles in the direction shown, along a rough table. The coefficient of friction between the particles and the table is $\frac{1}{4}$.

a Find:

   i the initial acceleration

   ii the tension in the string.

8 seconds later, particle B hits a wall and does not bounce back. The string goes slack.

b How much longer does it take for A to stop?

c How much further does A travel before it stops?

d If the string had been 2 m in length instead, what difference would this have made?

Solution

a i Label the forces on the diagram. There are two weights, two tensions, two reactions, two frictions and the 15 N force.

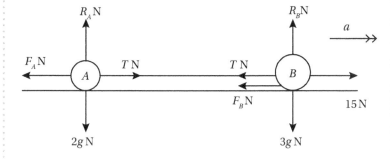

By resolving in the direction of motion for each particle in turn, you obtain simultaneous equations.

Resolve horizontally for B.

$R(\rightarrow)$

$F = ma$

$15 - T - F_B = 3a$ ①

Resolve horizontally for A.

$R(\rightarrow)$

$F = ma$

$T - F_A = 2a$ ②

Add equations ① and ② to eliminate T.

$15 - F_A - F_B = 5a$

Expressions for F_A and F_B can be determined by resolving vertically and applying $F = \mu R$. Note that there is no acceleration vertically.

Resolve vertically for B.

$R(\uparrow)$

$R_B - 3g = 0$

$R_B = 3g$ N

$F = \mu R$.

Hence $F_B = \frac{1}{4} \times 3g = \frac{3}{4} g$ N.

Similarly, $F_A = \frac{1}{4} \times 2g = \frac{1}{2} g$ N.

Therefore $15 - \frac{1}{2} g - \frac{3}{4} g = 5a$.

$5a = 15 - 5 - 7.5 = 2.5$

$a = 0.5 \, \text{m s}^{-2}$

The initial acceleration is $0.5 \, \text{m s}^{-2}$.

ii From equation ②, $T - F_A = 2a$.

$T - \frac{1}{2} g = 2 \times 0.5$

$T = 2 \times 0.5 + 5 = 6$ N

b During the first stage of the motion, $u = 0 \, \text{m s}^{-1}$, $a = 0.5 \, \text{m s}^{-2}$ and $t = 8$ s.

Substitute into $v = u + at$ to find the final velocity for the first stage (and initial velocity for the second stage).

$v = 0 + 0.5 \times 8$

$= 4 \, \text{m s}^{-1}$

Always resolve in the direction of motion when using $F = ma$ (when $a \neq 0 \, \text{m s}^{-2}$).

Once B hits the wall, the string will go slack. In order to answer the questions in **parts b, c** and **d**, it is necessary to consider the two stages of motion.

Make sure you can remember all of the equations of uniformly accelerated motion.

During the second stage of the motion for A, $u = 4\,\text{m}\,\text{s}^{-1}$ and $v = 0\,\text{m}\,\text{s}^{-1}$.

To find the acceleration for the second stage, resolve horizontally again for A but without the tension force.

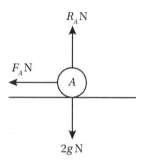

$R(\rightarrow)$

$F = ma$

$-F_A = 2a$

The reaction force has not changed so F_A is still

$\frac{1}{2}g\,\text{N} - \frac{1}{2}g = 2a$

$\quad a = -2.5\,\text{m}\,\text{s}^{-2}$

Substitute into $v = u + at$ to find the time taken for A to stop.

$t = \dfrac{v - u}{a}$

$\quad = \dfrac{0 - 4}{-2.5}$

$\quad = 1.6\,\text{s}$

c Substitute $u = 4\,\text{m}\,\text{s}^{-1}$, $v = 0\,\text{m}\,\text{s}^{-1}$ and $a = -2.5\,\text{m}\,\text{s}^{-2}$ into $v^2 = u^2 + 2as$ to find the displacement for the second stage.

$s = \dfrac{v^2 - u^2}{2a}$

$\quad = \dfrac{0 - 4^2}{2 \times -2.5}$

$\quad = 3.2\,\text{m}$

> Note that you could use $s = ut + \frac{1}{2}at^2$ but this assumes that the answer to **part b** for the time is correct (and accurate). It is usually better to use the information that you are given, where possible, rather than relying on a previous answer.

d If the string had only been 2 m in length, then A would have hit B and the wall since 2 m is less than the 3.2 m A would have travelled if the wall had not been there.

Now consider situations in which you need to resolve forces at angles into components, such as a car towing a trailer up a rough hill. The force diagram would look like this:

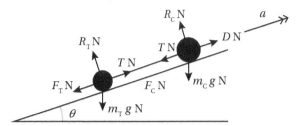

The main difference between this situation and that on a horizontal road is that the weight forces will need to be resolved into components.

Resolve for the car and trailer parallel to the slope.

R($\nearrow$)car

$$D - T - F_C - m_C g \sin\theta = m_C a$$

R($\nearrow$)trailer

$$T - F_T - m_T g \sin\theta = m_T a$$

Resolve for the car and trailer perpendicular to the slope.

R($\nwarrow$)car

$$R_C = m_C g \cos\theta$$

R($\nwarrow$)trailer

$$R_T = m_T g \cos\theta$$

Example 10

A van of mass 500 kg pulls a trailer of mass 200 kg up a hill inclined at 11° to the horizontal. The coefficient of friction between each vehicle and the hill is $\frac{3}{20}$. Find the driving force exerted by the engine if the van is accelerating at 1.5 m s⁻².

Solution

Start by drawing the force diagram.

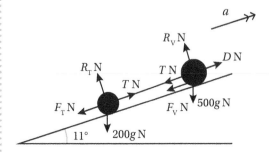

Resolve parallel to the hill for the van in the direction of motion.

$R(\nearrow)$

$F = ma$

$D - T - F_V - 500g\sin 11° = 500 \times 1.5$ ①

Resolve parallel to the hill for the trailer.

$R(\nearrow)$

$T - F_T - 200g\sin 11° = 200 \times 1.5$ ②

Add equations ① and ②.

$D - F_V - F_T - 500g\sin 11° - 200g\sin 11° = 700 \times 1.5$

Find expressions for F_A and F_B by resolving perpendicular to the hill and applying $F = \mu R$.

Resolve perpendicular to the hill for the van and trailer.

$R(\nwarrow)$

$R_V - 500g\cos 11° = 0$, so $F_V = \dfrac{3}{20} \times 500g\cos 11°$

$R_T - 200g\sin 11° = 0$, so $F_T = \dfrac{3}{20} \times 200g\cos 11°$

Hence:

$D - \dfrac{3}{20} \times 500g\cos 11° - \dfrac{3}{20} \times 200g\cos 11° - 500g\sin 11°$
$\quad - 200g\sin 11° = 700 \times 1.5$

$$D - 2366 = 1050$$

$$D = 3420\,\text{N (to 3 s.f.)}$$

The driving force exerted by the engine is 3420 N.

> Round answers to mechanics problems to 3 significant figures unless asked to do otherwise.

Exercise 4.4A

1. Two particles, G and H, are at rest on a smooth horizontal table and connected by a light inextensible string. H is given a force of 3 N, which causes the particles to move. The mass of G is 150 g and the mass of H is 350 g. Find the tension in the string and the acceleration of G.

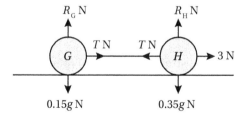

2. A particle P, of mass $4m$ kg, is suspended from a vertical rope. A particle Q, of mass $3m$ kg, is suspended from particle P by another vertical rope. A vertical force, X N, is applied to P such that the particles accelerate upwards at $\dfrac{4}{7}g$. Find the tension in the rope and the value of X, giving the answers in terms of m and g.

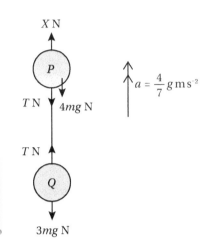

3 A car of mass 300 kg is towing a caravan of mass 840 kg along a straight horizontal stretch of motorway. The engine of the car is exerting a driving force of 1.71 kN. Both vehicles experience resistive forces that are proportional to their masses. Find the tension in the tow bar.

4 Two particles of masses 7 kg and 4 kg on a rough horizontal table are connected by a light inextensible 50 cm string. The heavier particle is projected away from the lighter particle by a force of 35 N. The numerical value of the coefficient of friction between each particle and the table is one-twentieth of its mass. After 4 seconds, the heavier particle hits a wall and does not bounce back.

   a Does the lighter particle collide with the heavier particle?

   b If it does, how much longer does it take the lighter particle to reach the heavier particle; if it does not, how far apart do they end up?

5 A particle, X, of mass 5 kg, is suspended from a vertical string. A second particle, Y, of mass 3 kg, is suspended from X by a vertical cord. A force of 100 N is applied to the string and the two particles accelerate upwards. When the particles are travelling at $9 \, \text{m s}^{-1}$, the cord snaps. How long does it take Y to return to its original position?

6 A minibus of mass 950 kg is towing a small trailer along a straight horizontal road. The resistances to motion are 475 N and 25 N, respectively. While the minibus is increasing in speed from $12 \, \text{m s}^{-1}$ to $32 \, \text{m s}^{-1}$ in 8 s, the driving force of the minibus is 3 kN. When the minibus reaches $32 \, \text{m s}^{-1}$, the minibus and trailer become uncoupled.

   a Find the mass of the trailer.

   b Show that the trailer travels 1024 m before it comes to rest.

7 A car of mass M kg is towing a caravan of mass 750 kg along a straight horizontal road. The resistance on the car is X N and the resistance on the caravan is 400 N. When the engine of the car exerts a driving force of 1450 N, the car accelerates at $0.6 \, \text{m s}^{-2}$ but when the engine of the car exerts a driving force that is 1 kN larger, the car accelerates at $1.4 \, \text{m s}^{-2}$.

   a Find the value of M.

   b Find the value of X.

8 Two particles, X and Y, are connected by a light string, with X vertically above Y. The masses of X and Y are 6 kg and 14 kg, respectively. X experiences a force of 208 N pulling it upwards for 5 s from rest. At this point, the string snaps. Find the further time it takes for Y to reach its maximum height.

9 A car of mass 900 kg pulls a caravan of mass 1200 kg up a rough hill inclined at 7° to the horizontal. The acceleration of the car as it drives up the hill is 0.65 m s^{-2}. After a certain time, the hill increases in steepness so that it is then inclined at 10° to the horizontal, and under these conditions the acceleration of the car reduces to 0.15 m s^{-2}. The driving force exerted by the engine of the car is the same for each gradient. The coefficient of friction between each vehicle and the hill is constant throughout.

  a Find the coefficient of friction.

  b Find the driving force exerted by the engine of the car.

10 A small train is made from three sections, A, B and C, such that the engine A pulls the two carriages B and C along a horizontal track. The mass of B is M kg. B is 40% lighter than A and 50% heavier than C. The coefficient of friction between each section and the track is $\frac{1}{10}$. The driving force exerted by the engine is 8.4 kN. Having travelled 2.79 km in 3 minutes at a constant acceleration and reaching a velocity of 20 m s^{-1}, C becomes uncoupled from the rest of the train.

  a Find the value of M.

  b Find the tension in the coupling between A and B immediately after C is uncoupled.

  c Find the time taken for C to stop after the uncoupling.

Mathematics in life and work: Group discussion

As part of your role in research and development at a car manufacturer, you are required to model the forces that will act upon the car and investigate how this will affect the motion of the car. Consider modelling a car towing a caravan up a hill.

1 Sketch a force diagram for this situation. What assumptions and/or simplifications have you made to produce your model?

An initial model for a car pulling a caravan has the following details. The car has a mass of 700 kg and its engine can produce a driving force of up to 1 kN. It needs to pull a caravan of mass 2000 kg. You are investigating this model for a car pulling the caravan up a hill inclined at an angle of 3° with the coefficient of friction between each vehicle and the hill equal to 0.05.

2 Find the acceleration of the car. What do you notice?

3 What modifications could you make to this situation to make the required motion possible?

Another situation in which objects are connected is when a person is travelling in a lift. In this situation, the connection is the contact between the person and the floor of the lift, as discussed in **Example 11**.

Example 11

A woman is standing in a lift that is being pulled upwards by a cable. The woman has a mass of 80 kg and the lift has a mass of 400 kg. Given that the reaction force, R, exerted on the woman by the lift is 920 N, find the tension in the cable.

Solution

First, represent the information in a diagram.

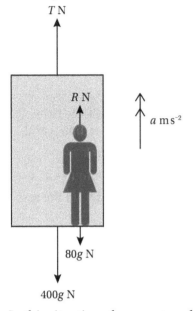

> ### KEY INFORMATION
> Problems involving lifts often require you to consider the reaction force acting on a person in the lift, as well as considering the lift as a whole.

In this situation, there are two different resultant forces, one for the system as a whole and one for the interaction between the woman and the floor of the lift.

Considering the woman and the floor, the resultant force is given by $(R - 80g)$ N, where $R = 920$ N.

Substitute the resultant force into $F = ma$.

$$920 - 80g = 80a$$
$$920 - 80 \times 10 = 80a$$
$$120 = 80a$$
$$a = 1.5\,\text{m}\,\text{s}^{-2}$$

Considering the system as a whole, the force downwards is provided by the weights of both the woman and the lift.

$$80g + 400g = 480g$$

The resultant force is given by $(T - 480g)$ N.

Substitute into $F = ma$.

$$T - 480g = 480 \times 1.5$$
$$T = 480g + 480 \times 1.5$$
$$T = 480(10 + 1.5)$$
$$T = 5520\,\text{N}$$

Exercise 4.4B

1 A teenager standing in a lift has a mass of 55 kg.

Find the reaction force exerted on the teenager by the lift when the lift is:

a accelerating upwards at $2.4\,\mathrm{m\,s^{-2}}$

b accelerating downwards at $1.6\,\mathrm{m\,s^{-2}}$

c decelerating upwards at $0.6\,\mathrm{m\,s^{-2}}$.

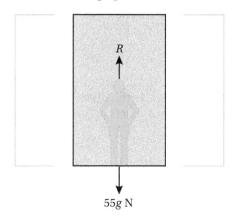

R

$55g$ N

2 A dog of mass 25 kg sits in a lift of mass 140 kg. The lift is descending at $1.8\,\mathrm{m\,s^{-2}}$.

a Find the reaction force exerted by the lift on the dog.

b Find the tension in the lift cable.

3 Two adults, of masses 75 kg and 65 kg, enter a lift. The lift has a mass of 550 kg. The tension in the lift cable is 7616 N. Find the time it takes for the lift to ascend 24 m.

4 Five people, of average mass 66 kg, travel upwards in a lift of mass 280 kg, at $0.2\,\mathrm{m\,s^{-2}}$.

a Find the tension in the lift cable.

When a sixth person joins the lift, the average mass increases by 4 kg.

Given that the acceleration decreases by $0.05\,\mathrm{m\,s^{-2}}$,

b find the reaction force exerted on the sixth person by the floor of the lift

c find the new tension in the lift cable.

5 A girl and her father stand in a lift which is travelling downwards at $\frac{1}{2}\,\mathrm{m\,s^{-2}}$. The father weighs twice as much as the girl and the lift weighs 100 kg more than the father. The tension in the lift cable is 2.66 kN. Find the reaction force exerted by the lift floor on the father.

6 Liu and Zhao, whose masses are in the ratio 2 : 3, respectively, take separate journeys in a lift of mass 360 kg. When Liu ascends at $a\,\mathrm{m\,s^{-2}}$, the tension in the lift cable is 4.73 kN. When Zhao descends at $(a - 0.25)\,\mathrm{m\,s^{-2}}$, the reaction force exerted on him by the lift is 1.14 kN.

a Find the value of a.

Zhao and Liu now take the lift at the same time, ascending at $0.6\,\mathrm{m\,s^{-2}}$.

b Find the tension in the lift cable.

PS **7** A lift of mass 400 kg has a notice stating that the maximum load is 160 kg. When the lift ascends with a load of this mass, the lift accelerates at 0.75 m s^{-2}. However, a caretaker carelessly puts a 200 kg crate in the lift whilst the lift is on the ground floor, without checking the notice. After the lift has been ascending for 30 s, the cable snaps and the lift returns to the ground floor. Assuming that the tension in the cable before it snaps is the same as when the load was 160 kg, how long after the cable snaps does it take for the lift to return to the ground floor?

PS **8** A hotel has a lift of mass 280 kg.

   a Margarita gets into the lift. As the lift ascends, the tension in the lift cable is 4320 N and Margarita experiences a reaction force of 960 N from the floor of the lift. Find:

     i Margarita's mass

     ii the acceleration of the lift.

   b Margarita gets out of the lift and Guillaume and Josef get in. Now, as the lift ascends, the tension in the lift cable is 5100 N. Guillaume and Josef experience reaction forces of 700 N and 900 N, respectively, from the floor of the lift. Find:

     i Guillaume's mass

     ii Josef's mass

     iii the acceleration of the lift.

4.5 Pulleys

Another example of a system with two connected particles is the **pulley** system. Consider two particles, A and B, of masses 5 kg and 10 kg, respectively, suspended vertically, as shown in the diagram, and connected by a string passing over a pulley.

Since B is heavier than A, B will pull A. Both particles will experience a resultant force and B will accelerate downwards whilst A accelerates upwards.

> **KEY INFORMATION**
> Objects connected via a pulley will travel in different directions.

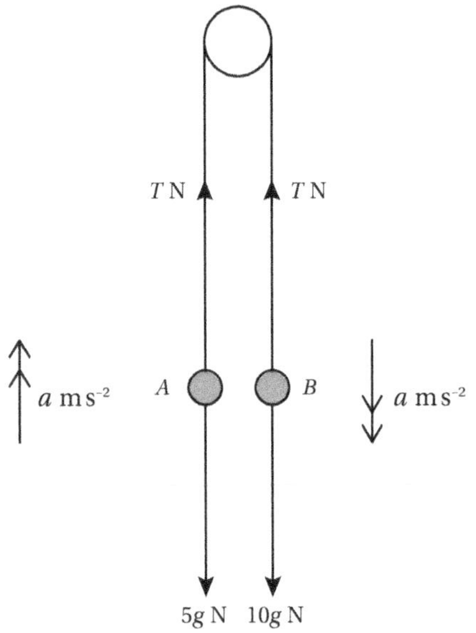

The main difference between objects connected via a pulley and connected objects such as cars and trailers is that objects connected via a pulley will be travelling in different directions.

Modelling assumptions

› Tensions are equal.

› Strings and ropes are modelled as light and inextensible.

› Pulleys are assumed to be fixed and smooth (i.e. without friction).

Example 12

Two particles, A and B, of masses 11 kg and 3 kg, respectively, are connected to either end of a light inextensible string via a smooth pulley, as shown in the diagram. Initially, the particles are at rest. Find the tension in the string and the acceleration of the two particles when the particles are released. Let $g = 10\,\mathrm{m\,s^{-2}}$.

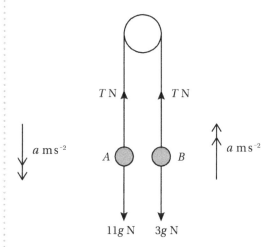

Solution

In this example, both the objects travel vertically, but one is going upwards and the other is going downwards.

Because particle A is heavier than particle B, particle A will move downwards and particle B will move upwards.

For particle A moving downwards, the resultant force $= (11g - T)\,\mathrm{N}$.

Substitute the resultant force into the formula $F = ma$.

$$11g - T = 11a \qquad\qquad ①$$

For particle B moving upwards, the resultant force $= (T - 3g)\,\mathrm{N}$.

> The resultant force for each particle will be in the direction that each particle is moving.

Substitute the resultant force into the formula $F = ma$.

$$T - 3g = 3a \qquad \qquad ②$$

Again, these are simultaneous equations. Adding them together will eliminate T.

$$11g - 3g = 11a + 3a$$

$$8g = 14a$$

$$a = \frac{8g}{14} = \frac{4g}{7}$$

$$= \frac{40}{7} \, \text{m s}^{-2}$$

Substitute $a = \frac{40}{7} \, \text{m s}^{-2}$ into equation ②.

$$T - 3g = 3a$$

$$T = 3g + 3a$$

$$= 3 \times 10 + 3 \times \frac{40}{7}$$

$$= \frac{330}{7} \, \text{N}$$

In **Example 12**, both objects travel vertically with the heavier particle, A, moving downwards and the lighter particle, B, moving upwards. In **Example 13**, although particle D travels vertically, particle C is on a horizontal table and is pulled along the table by particle D. Hence the weight of particle D will not feature in its resultant force.

Example 13

Particle C, of mass 4 kg, is connected to particle D, of mass 5 kg, by a light inextensible string that passes over a smooth pulley at the edge of a table. C is on the table 1.5 m from the pulley. D is suspended vertically below the pulley and 0.6 m above the floor. Both particles are held in position with the string taut until they are released. In the subsequent motion, D hits the floor and remains at rest on the ground and C experiences a resistive force of 23 N as it is pulled across the table.

Find the distance of C from the pulley when it comes to rest.

Solution

In this example, one object is suspended vertically from a pulley but the other is on a horizontal surface.

Start by putting the information in a diagram. Let the tension be T N, the acceleration be a m s^{-2} and the reaction force exerted on C by the table be R N.

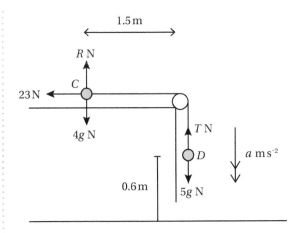

Since the question describes the motion of C and D, you can assume that the resistive force is not sufficient to prevent motion. Particle D will fall vertically, pulling C across the table.

For particle D, the resultant force $= (5g - T)\,\mathrm{N}$.

Substituting into $F = ma$ gives

$$5g - T = 5a \qquad \qquad ①$$

For particle C moving across the table, the resultant force $= (T - 23)\,\mathrm{N}$.

Substituting into $F = ma$ gives

$$T - 23 = 4a \qquad \qquad ②$$

Adding the simultaneous equations gives $5g - 23 = 9a$.

$$5 \times 10 - 23 = 9a$$

$$50 - 23 = 9a$$

$$9a = 27$$

$$a = 3\,\mathrm{m\,s^{-2}}$$

Particles C and D will both move with an acceleration of $3\,\mathrm{m\,s^{-2}}$ until D hits the floor, at which point the string will become slack and D will no longer be pulling C.

$$u = 0\,\mathrm{m\,s^{-1}}, s = 0.6\,\mathrm{m}, a = 3\,\mathrm{m\,s^{-2}}$$

$$v^2 = u^2 + 2as$$

$$= 0^2 + 2 \times 3 \times 0.6$$

$$= 3.6$$

$$v = 1.90\,\mathrm{m\,s^{-1}}$$

When the string becomes slack, C will be slowed down by the resistive force but there will be no tension force pulling it. Hence C will decelerate and come to rest.

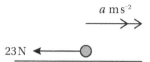

The new resultant force for particle $C = -23\,\text{N}$.

Substituting into $F = ma$ gives $-23 = 4a$.

$a = -5.75\,\text{m s}^{-2}$.

$u = 1.90\,\text{m s}^{-1}$, $v = 0\,\text{m}$, $a = -5.75\,\text{m s}^{-2}$

$v^2 = u^2 + 2as$

$s = \dfrac{v^2 - u^2}{2a}$

$= \dfrac{0^2 - 1.90^2}{2 \times -5.75}$

$= 0.313\,\text{m}$

Hence the total distance travelled by C is $0.6 + 0.314 = 0.914\,\text{m}$.

C is $1.5 - 0.914 = 0.587\,\text{m}$ from the pulley when it comes to rest.

> The initial velocity when the string is slack is the final velocity when D hits the floor.

Now you will consider a pulley system where either or both objects are on an inclined plane.

The force diagrams below show the difference between the situations for a smooth slope and a rough slope, where particle B, of mass $2m\,\text{kg}$, is suspended freely from a pulley and particle A, of mass $m\,\text{kg}$, is on the slope.

Smooth slope

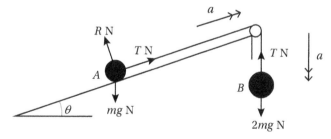

R($\nearrow$) for A

$T - mg\sin\theta = ma$ ①

R($\downarrow$) for B

$2mg - T = 2ma$ ②

Add equations ① and ②.

$2mg - mg\sin\theta = 3ma$

Rough slope

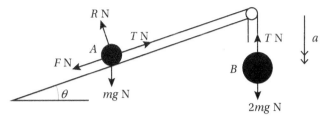

R($\nearrow$) for A

$T - mg\sin\theta - F = ma$ ①

R($\nwarrow$) for A

$R - mg\cos\theta = 0$

R($\downarrow$) for B

$2mg - T = 2ma$ ②

Add equations ① and ②.

$2mg - mg\sin\theta - F = 3ma$

$2mg - mg\sin\theta - \mu mg\cos\theta = 3ma$

Example 14

Particle A, of mass 2 kg, lies on a rough plane inclined at an angle θ to the horizontal, where $\cos\theta = \frac{8}{17}$. The coefficient of friction between A and the plane is $\frac{1}{4}$. Particle A is connected to particle B, of mass 3 kg, by a light inextensible string which lies along a line of greatest slope of the plane and passes over a smooth pulley. The system is initially held at rest with B hanging vertically 2 m above the ground.

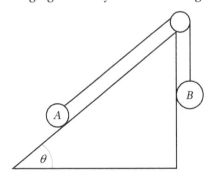

The system is released. Find:

a the initial acceleration of B

b the tension

c the total distance travelled by A before it comes to rest, assuming that B does not bounce when it hits the ground.

Solution

a Label the diagram with its six forces and two accelerations. Since only one particle lies on the slope, there is no need for subscripts.

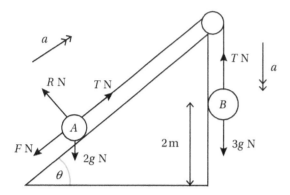

Resolve parallel to the plane in the direction of motion of A to find an equation in terms of T, F and a.

R($\nearrow$)

$$T - 2g\sin\theta - F = 2a \qquad \text{①}$$

Resolve vertically downwards in the direction of motion of B to find an equation in terms of T and a.

R($\downarrow$)

$$3g - T = 3a \qquad \text{②}$$

Add equations ① and ②.

$$3g - 2g\sin\theta - F = 5a$$

Resolve perpendicular to the plane for A to find the reaction force.

R($\nwarrow$)

$$R - 2g\cos\theta = 0$$
$$F = \mu R = \tfrac{1}{4} \times 2g\cos\theta$$

Hence

$$3g - 2g\sin\theta - \tfrac{1}{4} \times 2g\cos\theta = 5a$$

Note that if $\cos\theta = \frac{8}{17}$ and θ is acute, then $\sin\theta = \frac{15}{17}$.

Recall that this can be shown using Pythagoras' theorem, as follows.

Sketch a right-angled triangle with an adjacent side of 8 and a hypotenuse of 17.

By Pythagoras' theorem, $a^2 + 8^2 = 17^2$

$$a^2 + 64 = 289$$

$a^2 = 289 - 64 = 225$

$a = \sqrt{225} = 15$

Hence $\sin\theta = \dfrac{\text{opposite (opp)}}{\text{hypotenuse (hyp)}} = \dfrac{O}{H} = \dfrac{15}{17}$.

So $3g - 2g \times \dfrac{15}{17} - \dfrac{1}{4} \times 2g \times \dfrac{8}{17} = 5a$.

$3g - \dfrac{30}{17}g - \dfrac{4}{17}g = 5a$

$$5a = g$$

$$a = \tfrac{1}{5}g = 2\,\text{m s}^{-2}$$

The initial acceleration of B is $2\,\text{m s}^{-2}$.

b From equation (2), $3g - T = 3a$

$T = 3g - 3a$

$\quad = 3 \times 10 - 3 \times 2$

$\quad = 24$

The tension is $24\,\text{N}$.

c There are two stages to A's motion before it comes to rest – the first stage when it is pulled by B until B hits the ground and the second stage when it is slowed down to instantaneous rest by friction and gravity.

During the first stage, $a = 2\,\text{m s}^{-2}$, $s = 2\,\text{m}$ and $u = 0\,\text{m s}^{-1}$.

Substitute into $v^2 = u^2 + 2as$ to find the final velocity (and the initial velocity for the second stage).

$v^2 = 0 + 2 \times 2 \times 2$

$\quad = 8$

$v = \sqrt{8}\,\text{m s}^{-1}$

For the acceleration during the second stage, resolve again for A but without tension. The reaction, and hence the friction, will be unchanged.

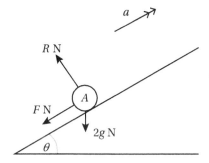

Draw a new force diagram because the situation has changed.

R($\nearrow$)

$-2g \times \dfrac{15}{17} - \dfrac{1}{4} \times 2g \times \dfrac{8}{17} = 2a$

$2a = -2g$

$a = -g = -10\,\text{m}\,\text{s}^{-2}$

So during the second stage, $a = -10\,\text{m}\,\text{s}^{-2}$, $u = \sqrt{8}\ \text{m}\,\text{s}^{-1}$ and $v = 0\,\text{m}\,\text{s}^{-1}$

Substitute into $v^2 = u^2 + 2as$ to find the displacement during the second stage.

$s = \dfrac{v^2 - u^2}{2a}$

$= \dfrac{0 - \left(\sqrt{8}\right)^2}{2 \times -10}$

$= 0.4\,\text{m}$

Combining the 2 m travelled during the first stage, the total distance travelled by A is 2.4 m.

> Make sure you read the question carefully so that you know whether you need the total distance or the further distance.

Exercise 4.5A

1 Particles G and H, of masses $3m$ kg and $2m$ kg, are connected by a light inextensible string with G at rest on a smooth table. The string passes over a smooth fixed pulley at the edge of the table with H hanging freely beneath the pulley.

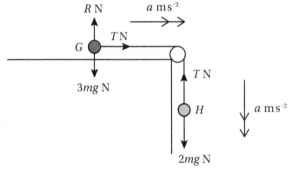

The system is released from rest.

a Find the force exerted on G by the table in terms of m and g.

b Find the acceleration of G (or of H) in terms of g.

c Find the tension in the string in terms of m and g.

2 Two particles, of masses 3 kg and 4 kg, respectively, are connected by an inextensible string via a smooth fixed pulley, both hanging vertically. The system is released from rest.

a Find the acceleration of either particle in terms of g.

b Find the tension in the string in terms of g.

c Find the difference in height between the particles after 1.4 s.

3 Particles P_1 and P_2 are connected by a light inextensible cord passing over a smooth fixed pulley. P_1 has a mass of 4 kg and P_2 is heavier than P_1. When the system is released from rest, the acceleration of each particle is equal to $\frac{1}{5}g$ m s^{-2}.

   a Find the mass of P_2.

   b Find the force exerted on the pulley.

4 A particle J sits on a rough horizontal table connected by a light inextensible string via a smooth fixed pulley at the edge of the table to a particle K, which is suspended 3 m above the ground. Particles J and K have masses 2 kg and 3.2 kg, respectively. The coefficient of friction between particle J and the table is 0.3. K is released from rest with the string taut.

   a Find the friction experienced by particle J.

   b Find the acceleration of J and K.

   c Find the tension in the string.

   d Assuming that J does not meet the pulley, how far does it travel altogether before it stops?

5 Particles A and B, of masses 3 kg and 7 kg, are connected by a light inextensible wire that passes over a smooth pulley. A is at rest on a smooth table 5 m from the pulley while B is hanging freely 2 m above the ground. When the system is released, how long does it take A to reach the pulley?

6 Two particles of masses x kg and y kg are connected by a light inextensible string passing over a smooth fixed pulley. Both particles are hanging freely. Show that when the system is released from rest, the tension is given by $\dfrac{2xyg}{x+y}$.

7 A ball of mass 500 g is suspended vertically from a smooth fixed pulley at the edge of a rough table. The ball is connected via a light inextensible string and the pulley to a flowerpot of mass 300 g on the table. The coefficient of friction between the flowerpot and the table is $\frac{2}{3}$. The ball is released and the string snaps after 0.2 s. Assuming that the flowerpot does not reach the pulley, find the total time for which the flowerpot is in motion.

8 Two particles, X and Y, of masses 700 g and 800 g, respectively, are connected by a light inextensible cord. X is at rest on a smooth slope inclined at an angle θ to the horizontal, where $\sin \theta = \frac{1}{4}$. The cord passes over a smooth light pulley fixed at the top of the slope. Y hangs freely from the pulley.

The system is released from rest with the string taut.

Calculate:

   a the acceleration of Y

   b the tension in the cord.

Y hits the ground after 3 s and remains at rest on the ground without bouncing.

   c Find the distance travelled by X before Y hits the ground.

In the subsequent motion, X does not reach the pulley.

   d Find the further distance travelled by X before it comes to instantaneous rest.

(c) **9** Two rocks, P and Q, are connected by an inextensible rope passing over a smooth pulley. Q is 3 kg heavier than P. Given that the tension in the rope is 45.5 N, find the mass of each rock and the acceleration of the system.

(c) **10** Two particles, A and B, of masses 1.8 kg and 5.2 kg, respectively, are connected by a light inextensible string passing over a fixed smooth light pulley. Particle A is placed on a rough plane inclined at 30° to the horizontal. The coefficient of friction between particle A and the plane is $\frac{\sqrt{3}}{9}$. Particle B hangs freely below the pulley. The particles are released from rest with the string taut.

a Show that the acceleration of the particles when released is given by $\frac{4}{7}g\,\mathrm{m\,s^{-2}}$.

b Calculate the tension in the string.

Particle B strikes the floor after 2 s.

c Find the total time from when the particles were released until A comes to instantaneous rest (assuming that it does not reach the pulley).

> **Stop and think**   In **question 4**, J is heavier than K so J moves down while K moves up. However, in **question 1**, H is lighter than G and yet is still able to pull it. How is this possible?

SUMMARY OF KEY POINTS

› Newton's three laws of motion:

  › Newton's first law states that a particle will remain at rest or continue to move with a constant velocity in a straight line unless acted upon by a resultant force.

  › Newton's second law states that the resultant force F acting upon a particle is proportional to the mass m of the particle and the acceleration a of the particle such that $F = ma$.

  › Newton's third law states that every action has an equal and opposite reaction.

› Weight, $W = mg$. Usually, g is taken as $10\,\text{m s}^{-2}$.

› The resultant force should be found in the direction of motion.

› An object is modelled as a particle so its mass is assumed to be concentrated at a single point.

› String is modelled as light and inextensible:

  › The string is modelled as light so that the tension is the same for both parts of the string.

  › The string is modelled as inextensible so that the acceleration is the same for both particles.

› Pulleys are modelled as smooth.

› Problems involving lifts often require you to consider the reaction force acting on a person in the lift as well as considering the lift as a whole.

› Problems involving connected particles usually require the solution of simultaneous equations.

› If two particles are connected by a string, then the tensions are equal.

› Particles connected via a pulley will be travelling in different directions.

EXAM-STYLE QUESTIONS

1   A box of mass M kg is pushed along a rough horizontal table by a horizontal force of 24 N. The coefficient of friction between the box and the table is 0.4.

   a Find the magnitude of the frictional force in terms of M and g.

   The box is accelerating at $2\,\text{m s}^{-2}$.

   b Write an equation of motion for the box.

   c Find the value of M.

2   A girl is travelling down a rough slope in a go-kart. The combined mass of the girl and the go-kart is 50 kg. The slope is inclined at an angle of 30° to the horizontal. The coefficient of friction between the go-kart and the slope is $\frac{1}{6}$.

   a Find the magnitude of the reaction force exerted on the girl and the go-kart by the slope.

   b Find the acceleration of the go-kart down the slope.

   c Given that the go-kart started from rest, how long does it take the girl to travel 15 m down the slope?

3 A particle A, of mass 5 kg, rests on a smooth horizontal table. Particle A is attached to one end of a light inextensible string which passes over a smooth pulley fixed at the edge of the table. The other end of the string is attached to a particle B, of mass 3 kg, which hangs freely below the pulley, 1.3 m above the ground. The system is released from rest with the string taut. In the subsequent motion, particle A does not reach the pulley before B reaches the ground.

a Find the tension in the string before B reaches the ground.

b Find the time taken by B to reach the ground.

4 A student is set the following question.

Two particles are attached to the ends of a light inextensible string which passes over a smooth fixed pulley. The particles are held in position with the string taut and the hanging parts of the string vertical. The particles are then released from rest. The particles have masses of 6 kg and m kg, where $m < 6$, and the initial acceleration of each particle has a magnitude of $\frac{2}{3}g$ m s^{-2}.

6 kg m kg

Find:

i the tension in the string immediately after the particles are released

ii the value of m.

The student's solution is shown below.

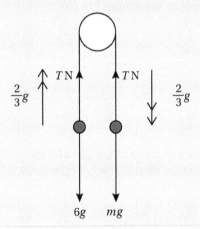

i Left:

$$T - 6g = 6 \times \tfrac{2}{3}g = 4g$$
$$T = 10g = 100\,\text{N}$$

ii Right:

$$mg - T = \tfrac{2}{3}mg$$
$$mg - 100 = \tfrac{2}{3}mg$$
$$\tfrac{1}{3}mg = 100$$
$$mg = 300$$
$$m = 300 \div 10 = 30\,\text{kg}$$

Assess the student's solution and make corrections where necessary.

5 Two particles, A and B, are connected by a light inextensible vertical cable with A above B. The mass of A is 500 g and the mass of B is 200 g. Initially, the system is at rest. A vertical force of X N is applied to A, which raises the particles 24 m in 4 s. Find the magnitude of X.

6 A man uses the lift in a hotel. The lift starts from rest on the ground floor and moves vertically upwards. The lift initially accelerates with a constant acceleration of $0.8\,\text{m s}^{-2}$ for 15 s. It then decelerates for 10 s before coming to rest.

The mass of the man is 90 kg and the mass of the lift is 160 kg. The lift is pulled up by means of a vertical cable attached to the top of the lift. By modelling the cable as light and inextensible, find:

a the tension in the cable when the lift is accelerating

b the magnitude of the force exerted by the lift on the man when the lift is decelerating.

c Explain how you have used the assumption that the cable is light.

7 A car of mass 400 kg pulls a horsebox of mass 600 kg up a rough hill inclined at $\sin^{-1} 0.2$. The resistances to motion are 400 N for the car and 600 N for the horsebox. The car exerts a driving force of 3.2 kN.

a Find the acceleration.

The car becomes uncoupled from the horsebox after travelling 90 m.

b Find the new acceleration of the car.

c Show that the horsebox travels 6 m after the uncoupling.

8 A particle P, of mass 7 kg, is projected vertically upwards from a point A with speed $22\,\text{m s}^{-1}$. The point A is 12.25 m above horizontal ground. The particle P moves freely under gravity until it reaches the ground with speed $V\,\text{m s}^{-1}$.

a Find the value of V.

The ground is soft and, after it reaches the ground, P sinks vertically downwards into the ground before coming to rest. The ground exerts a constant resistive force of magnitude 3000 N on P.

b Find the vertical distance that P sinks into the ground before coming to rest.

9 Two particles, P_1 and P_2, of mass 4.5 kg and 5.5 kg, respectively, initially at rest, are joined by a light horizontal rod, as shown in the diagram. A constant force of magnitude 7 N is applied to P_2 and the system moves under the action of this force for 10 s. During the motion, the resistance to the motion of P_1 has a constant magnitude of 2 N and the resistance to the motion of P_2 has a constant magnitude of 3 N.

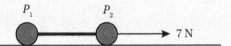

Find:

a the acceleration of the particles as the system moves under the action of the 7 N force

b the speed of the particles after 10 s

c the tension in the rod as the system moves under the action of the 7 N force.

After 10 s, the 7 N force is removed and the system decelerates to rest. The resistances to motion are unchanged.

Find:

d the distance moved by P_1 as the system decelerates

e the thrust in the rod as the system decelerates.

10 A van pulls a cart along a rough horizontal road. The mass of the van is 900 kg and the mass of the cart is 500 kg. The van and cart experience frictional forces of 800 N and 400 N, respectively. The tow bar coupling the van and cart together is inclined at 12° to the horizontal. The van exerts a driving force of 2.25 kN.

a Find the acceleration of the van.

b Show that the magnitude of the tension in the tow bar is 790 N, correct to 2 significant figures.

11 A car of mass 750 kg is pulling a trailer of mass 250 kg along a straight horizontal road by means of a tow bar. The resistances to motion of the car and the trailer are 300 N and 100 N, respectively. The engine of the car is exerting a constant driving force of 1.5 kN. The car and trailer are modelled as particles.

a Find the acceleration of the car and trailer.

b Find the magnitude of the tension in the tow bar.

c How have you used the assumption that the car and trailer can be modelled as particles?

The car is moving along the road when the driver sees a hazard ahead. He reduces the force produced by the engine to zero and applies the brakes. The brakes produce a force on the car of magnitude F N and both the car and trailer decelerate.

d Given that the resistances to motion are unchanged and that the magnitude of the thrust in the tow bar is 125 N, find the value of F.

12 Two particles, *A* and *B*, are such that *A* is heavier and has a mass of 2.8 kg. The particles are connected by a light inextensible string which passes over a smooth, fixed pulley. Initially, particle *A* is 3.75 m above a horizontal surface. The particles are released from rest with the string taut and the hanging parts of the string vertical. After particle *A* has been descending for 2.5 s, it strikes the ground. Particle *A* reaches the ground before particle *B* has reached the pulley.

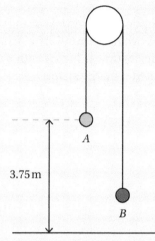

a Show that the acceleration of *A* as it descends is 1.2 m s⁻².

b Show that the mass of particle *B* is 2.2 kg.

c State how you have used the information that the string is inextensible.

When particle *A* strikes the ground, it does not rebound and the string becomes slack. Particle *B* then moves freely under gravity, without reaching the pulley, until the string becomes taut again.

d Find the time between the instant when particle *A* strikes the ground and the instant when the string becomes taut again.

13 A parcel *C*, of mass 2 kg, is at rest on a smooth slope inclined at an angle θ to the horizontal, where $\tan \theta = 0.75$. *C* is connected to another parcel *D*, of mass 3 kg, by means of a light inextensible rope and a smooth pulley at the top of the slope. *D* is suspended above the ground.

a How are *C* and *D* modelled? Explain your answer.

b Find the initial acceleration and tension.

After two seconds the rope snaps.

c How long does it take *C* to return to its original position after the rope snaps?

14

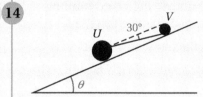

A particle *U* slides 32 m down a rough plane inclined at an angle θ to the horizontal, where $\cos \theta = \dfrac{24}{25}$, pulling particle *V* to which it is connected by a light string, as shown in the diagram. Particle *U* has a mass of *M* kg. The masses of *U* and *V* are in the ratio 5 : 2. Due to the relative sizes of the particles, the string makes an angle of 30° with the plane. The particles accelerate at $\frac{43}{625}$ g m s^{-2}. The coefficient of friction between U and the plane is 0.2.

a Find the coefficient of friction between V and the plane.

b Show that the tension in the string is given by $\frac{8\sqrt{3}}{625} Mg$.

15 Particle A is suspended from a vertical string and particle B is suspended by a vertical string from particle A. Initially, the system is at rest with particle B $2h$ m above the ground. The mass of particle B is 80% of the mass of particle A. A force is applied vertically upwards to A with a magnitude three times greater than the weight of B. After being pulled h m, the string between A and B breaks.

Show that the velocity of B when it returns to the ground is given by $v^2 = \frac{20gh}{3}$.

16

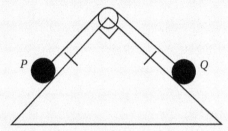

Blocks P and Q are connected by a light rope passing over a smooth pulley, as shown in the diagram. Each block is at rest on either side of a wedge. P has mass M kg and sits on the smooth side. Q is 5 kg heavier than P and sits on the rough side. The coefficient of friction between Q and the wedge is 0.2. The wedge is an isosceles triangle with a right angle at the top.

a Given that P accelerates at $2\sqrt{2}$ m s^{-2}, find the value of M.

b Find the force exerted on the pulley by the rope.

17

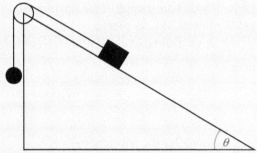

A block of mass 6.5 kg is connected to a ball of mass M kg by means of a light cord running across a smooth pulley. The block is at rest on a rough plane inclined at angle θ to the horizontal, where $\sin\theta = \frac{5}{13}$. The coefficient of friction between the block and the plane is $\frac{1}{3}$. The ball is suspended below the pulley, as shown in the diagram. A force of 7g N is applied to the block down the plane along the line of greatest slope.

a Given that the block moves down the slope, find the maximum value of M.

b If the ball is 3 kg lighter than the block, find the time taken for them to move 80 cm.

18 Particles P and Q, of masses 2.1 kg and 1.5 kg, respectively, are attached to the ends of a light inextensible taut string which passes over a smooth pulley fixed at the edge of a rough horizontal table. Initially, P is on the table 2.5 m from the pulley and Q hangs freely below the pulley, 1 m above the ground. The coefficient of friction between P and the table is given by μ. The system is released from rest.

   a Show that the initial acceleration of P is given by $\dfrac{(5 - 7\mu)g}{12}\,\text{m s}^{-2}$.

After Q hits the ground, P continues to move on the table, coming to rest before reaching the pulley.

   b Show that $\frac{1}{5} < \mu < \frac{5}{7}$.

19 A car is towing a milk float up a hill inclined at angle β to the horizontal. The masses of the car and milk float are 400 kg and 600 kg, respectively. The engine of the car exerts a driving force of 3.2 kN. Both vehicles experience a frictional force of $100g$ N and accelerate at 0.2 m s^{-2}.

Find:

   a the size of β in degrees, correct to 1 decimal place

   b the coefficient of friction between the car and the road.

Mathematics in life and work

A car manufacturer is using a software package to model the motion of a car towing a trailer up a plane. In the first scenario, the car is at rest in limiting equilibrium on a plane inclined at an angle of $\sin^{-1}\left(\dfrac{9}{41}\right)$ to the horizontal, with no trailer.

1 Find the coefficient of friction between the car and the plane.

In the second scenario, the car has been given a mass of 1230 kg and the car is to accelerate up the plane at 0.5 m s^{-2}.

2 Find the driving force required.

In the final scenario, the car is to tow a trailer of mass 82 kg up the plane. In the model, the coefficient of friction between the trailer and the plane is the same as between the car and the plane, and the driving force exerted by the engine of the car is the same.

3 Find the new acceleration of the car.

4 Comment on the model in the final scenario.

5 ENERGY, WORK AND POWER

Mathematics in life and work

The words work, energy and power are used in everyday language, for example: 'I have a lot of work to do today', 'I don't have enough energy to do that' or 'She's so powerful'. In Mathematics and Physics, these words have specific meanings that are similar to their common usage. Work and energy relate to forces acting over distances and power is the rate at which something is working.

There are many careers where an understanding of one or more of these concepts is necessary – for example:

» If you were an athlete preparing for a race, you would plan your diet to ensure that you were taking on an appropriate amount of chemical energy to be converted into work done against forces and kinetic energy. If your race involved jumping over hurdles, you would also need to consider potential energy.

» If you were designing a crane to lift and lower heavy objects, you would need to consider the kinetic and potential energy of the objects as they were moving to make sure that the crane would be able to move them safely.

» If you were designing a racing car, you would consider the power that could be developed by the engine of the car and how to reduce the resistance in order to maximise the speed that the car could attain during a race.

In this chapter, you will be looking at some of the factors to consider when designing a racing car.

LEARNING OBJECTIVES

You will learn how to:

» understand the concept of the work done by a force, and calculate the work done by a constant force when its point of application undergoes a displacement that is not necessarily parallel to the force

» understand the concepts of gravitational potential energy and kinetic energy, and use appropriate formulae

» understand and use the relationship between the change in energy of a system and the work done by the external forces, and use, in appropriate cases, the principle of conservation of energy

» use the definition of power as the rate at which a force does work, and use the relationship between power, force and velocity for a force acting in the direction of motion

» solve problems involving, for example, the instantaneous acceleration of a car moving on a hill against a resistance.

LANGUAGE OF MATHEMATICS

Key words and phrases you will meet in this chapter:

conservation of energy, energy, joule, kinetic energy, potential energy, power, watt, work done

PREREQUISITE KNOWLEDGE

You should already know how to:

> apply Pythagoras' theorem and the sine, cosine and tangent ratios for acute angles to the calculation of a side or an angle of a right-angled triangle

> use appropriate formulae for motion with constant acceleration in a straight line

> understand the vector nature of a force, and find and use components and resultants

> understand the concepts of limiting friction and limiting equilibrium, recall the definition of coefficient of friction and use the relationship $F = \mu R$ or $F \leqslant \mu R$, as appropriate

> apply Newton's laws of motion to the linear motion of a particle of constant mass moving under the action of constant forces, which may include friction, tension in an inextensible string and thrust in a connecting rod

> use the relationship between mass and weight

> solve simple problems that may be modelled as the motion of a particle moving vertically or on an inclined plane with constant acceleration

> solve simple problems which may be modelled as the motion of connected particles.

You should be able to complete the following questions correctly:

1 Find the height of this triangle, correct to 2 decimal places.

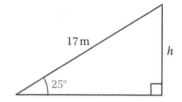

2 Find the time taken for a ball to drop 60 m vertically from rest.

3 A particle of mass 3.5 kg is pushed up a rough slope inclined at an angle of 33° to the horizontal by a force P N which is parallel to the slope. The coefficient of friction between the particle and the slope is $\frac{7}{10}$. Given that the particle moves at a steady speed, find the magnitude of P.

5.1 Work

Pushing a trolley requires you to use an amount of effort, and it requires more effort if the floor is rough or if you are pushing the trolley uphill. This is because you are working against forces – friction when pushing the trolley along a rough floor, and the weight of the trolley when pushing the trolley uphill. Also, the further you push the trolley the more work you have to do.

This combination of force and distance is called the **work done (WD)**. The work done against a force of F N for a distance of d m is given by the formula WD = Fd, where WD is measured in **joules** (J). For example, the work required to push an object with a force of 3 N for 4 m is $3 \times 4 = 12$ J. The units of WD can also be written as N m or as kg m^2 s^{-2}.

This formula is correct if the trolley is pushed horizontally along a horizontal floor or parallel to the slope. However, if the force is applied at an angle θ to the surface, as shown in the diagram, then you need to use the horizontal component of the force. In this case, the work done is given by $Fd \cos \theta$. For example, the work required to push an object with a force of 3 N at an angle of 20° for 4 m is $3 \times 4 \cos 20° = 11.3$ J.

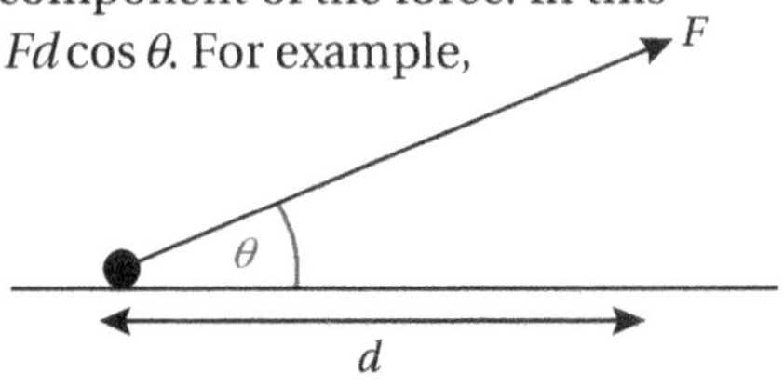

KEY INFORMATION

The formula for work done by or against a force is WD = Fd.

KEY INFORMATION

The formula for work done by or against a force acting at an angle of θ is WD = $Fd \cos \theta$.

Example 1

A vase is pushed 1.6 m across a horizontal table by a horizontal force of 8 N.

a Find the work done by the force in pushing the vase.

b Find the work done if the force acts at an angle of 30° to the horizontal instead.

Solution

a Work done is given by the formula WD = Fd.
WD = $8 \times 1.6 = 12.8$ J

b When the force acts at an angle θ, work done is given by the formula WD = $Fd \cos \theta$.
WD = $8 \times 1.6 \times \cos 30° = 11.1$ J

Example 2

A spanner is pushed 0.4 m across a rough horizontal table. The spanner experiences a frictional force of 3 N. Find the work done against friction to push the spanner.

Solution

Use the formula WD = Fd.

WD = $3 \times 0.4 = 1.2$ J

Example 3

A case of mass 3.5 kg is lifted vertically through a distance of 1.8 m.

Find the work done against gravity to lift the case.

Solution

In this situation, the force is the weight of the case.

This force is given by $W = mg$ (where W is the weight of the case in newtons).

$W = 3.5 \times 10 = 35\,\text{N}$

Use the formula $WD = Fd$ (where W is the work done).

$WD = 35 \times 1.8 = 63\,\text{J}$

Recall that $g = 10\,\text{m}\,\text{s}^{-2}$ unless stated otherwise.

Note that the work done against gravity is often rewritten as $WD = mgh$ (where mg is the weight force and h is the vertical distance (height)). This formula will be used for potential energy in **Section 5.2**.

Exercise 5.1A

1   Find the work done by:

   a   a horizontal force of 4 N pushing a lamp 2.1 m across a table

   b   a horizontal force of 5.2 kN pushing a car 1.3 m across a road

   c   a horizontal force of 9.3 N pushing a lawnmower 80 cm across a lawn.

2   Find the work done by:

   a   a 6 N force acting at 18° pushing a bag 4 m across a floor

   b   a 17 N force acting at an angle of $\cos^{-1}\left(\frac{4}{5}\right)$ pushing a hot water bottle 40 cm across a table

   c   a 380 kN force acting at an angle of $\tan^{-1}\left(\frac{12}{35}\right)$ pushing a truck 7.4 m across a car park.

3   **a**   An armchair is pushed 2.5 m across a floor. Given that the work done by the force is 400 J, find the force applied to the armchair.

   b   A sofa is pushed by a force of 750 N across a floor. Given that the work done by the force is 1.05 kJ, find the distance moved by the sofa.

4   A crate of bricks is lifted through a vertical distance of 18 m. The crate of bricks has a mass of 900 kg. Find the work done against gravity.

5   A refrigerator of mass 130 kg is pulled across a rough kitchen floor. The magnitude of the frictional force is 440 N. Given that the work done against friction is 528 J, find the distance the refrigerator is pulled.

6   A table is raised vertically 5.4 m to the second floor of a house. Given that the work done against gravity is 2295 J, find the mass of the table.

7   A van of weight 16 kN is lowered through a vertical distance of 6.5 m. Find the work done by gravity in lowering the van.

Example 4

A package of mass 23 kg is pulled 5 m at a constant speed across a rough horizontal floor in an airport by a cord at an angle of $\sin^{-1}\left(\frac{7}{25}\right)$ above the horizontal. Given that the work done against friction is 240 J, find the coefficient of friction between the package and the floor.

Solution

Start by drawing a diagram. Let the angle of the cord be θ.

Hence $\sin\theta = \frac{7}{25}$.

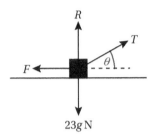

The work done against the frictional force, WD = Fd.

$$F = \frac{\text{WD}}{d} = \frac{240}{5} = 48\,\text{N}$$

Resolve horizontally in the direction of motion.

$R(\rightarrow)$

$$T\cos\theta - F = 0$$

$$F = T\cos\theta$$

Given that $\sin\theta = \frac{7}{25}$, use trigonometry to find the value of $\cos\theta$.

The adjacent is $\sqrt{25^2 - 7^2} = 24$.

Hence $\cos\theta = \frac{24}{25}$.

$$T = \frac{F}{\cos\theta} = \frac{48}{\frac{24}{25}} = 50\,\text{N}$$

Resolve vertically.

$R(\uparrow)$

$$R + T\sin\theta - mg = 0$$

$$R = mg - T\sin\theta = 23 \times 10 - 50 \times \frac{7}{25} = 230 - 14 = 216\,\text{N}$$

Finally, apply the formula $F = \mu R$.

$$\mu = \frac{F}{R} = \frac{48}{216} = \frac{2}{9}$$

Example 5

A cart of mass 8 kg is pushed 15 m up a rough plane inclined at an angle of 14°. The coefficient of friction between the cart and the plane is $\frac{1}{16}$. Find the total work done.

Solution

Start by drawing a diagram.

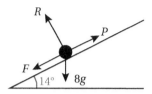

Work is done against both friction and gravity.

For the work done against friction, you need to know the frictional force. Find R by resolving perpendicular to the surface, then apply $F = \mu R$.

R(↖)

$R = mg\cos\theta$

Using $F = \mu R$, $F = \mu mg\cos\theta$.

$F = \frac{1}{16} \times 8 \times 10 \times \cos 14° = 4.851\,\text{N}$

WD $= Fd = 4.85 \times 15 = 72.77\,\text{J}$

For the work done against gravity, you need to know the height.

The height can be found by using trigonometry.

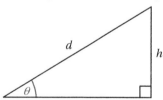

h is the opposite and d is the hypotenuse. Hence $h = d\sin\theta$.

$h = d\sin\theta = 15\sin 14° = 3.629$

WD $= Fd$

$\qquad = 8 \times 10 \times 3.629$

$\qquad = 290.3\,\text{J}$

Total work done $= 72.77 + 290.3 = 363\,\text{J}$, correct to 3 s.f.

KEY INFORMATION

For a surface inclined at an angle of θ to the horizontal, a distance of d m up the surface corresponds to a vertical distance given by $h = d\sin\theta$.

Exercise 5.1B

1 A trolley of mass 16 kg is pushed at a constant speed of $2.5\,\mathrm{m\,s^{-1}}$ for 8 s across a rough horizontal floor.

   a Find the distance travelled by the trolley.

   The force pushing the trolley, $X\,$N, is applied at an angle of 6° below the horizontal.

   b Find an expression for the friction force in terms of X.

   c Find an expression for the normal reaction force in terms of X.

   The coefficient of friction between the trolley and the floor is 0.3.

   d Find the value of X.

   e Find the work done by the force.

2 A shed is pushed 12.2 m up a rough plane inclined at $\cos^{-1}\left(\dfrac{60}{61}\right)$ to the horizontal by a force acting parallel to the plane. The mass of the shed is 244 kg. The coefficient of friction between the shed and the plane is $\dfrac{2}{15}$.

   a Find the magnitude of the force exerted by the plane on the shed.

   b Find the magnitude of the frictional force.

   c Find the work done against friction.

   d Find the vertical distance moved by the shed.

   e Find the work done against gravity.

   f Find the total work done moving the shed, in kJ correct to 3 significant figures.

3 A plane is inclined at an angle of 8.3° to the horizontal. A block of mass 9.5 kg is pulled 3.8 m up the plane. The plane is modelled as a rough surface and the block is modelled as a particle. The coefficient of friction between the block and plane is 0.22.

   a Find the work done against friction.

   b Find the work done against gravity.

4 A wheelbarrow of mass 25 kg is pushed up a rough track inclined at an angle of 30° to the horizontal at a speed of $0.2\,\mathrm{m\,s^{-1}}$.

   Given that the coefficient of friction between the wheelbarrow and the track is $\dfrac{\sqrt{3}}{15}$, find the total work done in one minute.

5 A particle of mass 3.6 kg is pulled 4.2 m at a constant speed up a rough inclined plane. The total work done is 43 J and the plane is inclined at an angle of 10° to the horizontal. Find the coefficient of friction between the particle and the plane.

6 A box of mass 6 kg is pushed up a rough driveway inclined at an angle of $\tan^{-1}\left(\dfrac{3}{4}\right)$ to the horizontal. The coefficient of friction between the box and the driveway is $\dfrac{3}{8}$. Given that the total work done is 513 J, find the distance the box is pushed up the driveway.

7 A hill is inclined at an angle of α to the horizontal. The hill is modelled as a rough plane with a coefficient of friction of 0.24. A young boy on a toboggan is pushed 15 m up the hill. The total mass of the boy and the toboggan is 51 kg. Given that the work done against gravity is 3.6 kJ, find:

   a the value of $\sin\alpha$

   b the work done against friction.

5.2 Kinetic energy and potential energy

There are various types of **energy** with which you will be familiar, such as chemical, electrical, heat, sound, light and nuclear. This section focuses on two types of energy that are properties of the motion and height of a particle: **kinetic energy** and **potential energy**.

Kinetic energy

Kinetic energy is the energy a particle has as a result of its motion.

In **Section 5.1** you saw that

work done = force × distance

From the equations of uniformly accelerated motion, distance is represented by the letter s.

work done = Fs

You also know that the resultant force is given by $F = ma$. Hence

work done = mas.

Rearrange $v^2 = u^2 + 2as$ to substitute for as.

$$2as = v^2 - u^2$$
$$as = \frac{1}{2}(v^2 - u^2)$$

Hence

work done = $m \times \frac{1}{2}(v^2 - u^2)$

$$= \frac{1}{2}m(v^2 - u^2).$$

The initial kinetic energy is given by $\frac{1}{2}mu^2$ and the final kinetic energy is given by $\frac{1}{2}mv^2$. The work done is the increase in kinetic energy. If $\frac{1}{2}m(v^2 - u^2)$ is positive, then the kinetic energy has increased, but if it is negative, then it has decreased.

Note that the units of kinetic and potential energy are the same as for work done: J, N m or kg m^2 s^{-2}.

> **KEY INFORMATION**
>
> The formula for kinetic energy for a particle of mass m kg moving at v m s^{-1} is
> $KE = \frac{1}{2}mv^2$.

> **KEY INFORMATION**
>
> The formula for the change in kinetic energy is
> $\frac{1}{2}m(v^2 - u^2)$.

Example 6

The initial velocity of a buggy of mass 25 kg is 28 m s^{-1}. Find the change in kinetic energy of the buggy after it decelerates at 1.6 m s^{-2} for 5 s, stating whether it is an increase or a decrease.

Solution

The formula for change in kinetic energy is given by

change in KE = $\frac{1}{2}m(v^2 - u^2)$.

You know that $m = 25$ kg and $u = 28$ m s^{-1} but you do not know the value of v.

You are given that $u = 28$ m s^{-1}, $a = -1.6$ m s^{-2} and $t = 5$ s.

Note that $m(v^2 - u^2)$ is not the same as $m(v - u)^2$.

Since you need v, you can substitute in the formula $v = u + at$.

$v = u + at$

$= 28 - 1.6 \times 5 = 20 \, \text{m} \, \text{s}^{-1}$

Change in KE $= \frac{1}{2} m(v^2 - u^2) = \frac{1}{2} \times 25 \times (20^2 - 28^2)$

$= \frac{1}{2} \times 25 \times (400 - 784)$

$= \frac{1}{2} \times 25 \times -384$

$= -4800 \, \text{J}$

Remember that energy is measured in joules (J).

This is a decrease of 4800 J, which could also be written as 4.8 kJ.

Because the answer is negative, the kinetic energy has decreased.

Example 7

Two particles, A and B, approach each other at $2 \, \text{m} \, \text{s}^{-1}$ and $3 \, \text{m} \, \text{s}^{-1}$ respectively. The mass of A is 4 kg and the mass of B is 2 kg. After the particles collide, A rebounds at $1 \, \text{m} \, \text{s}^{-1}$.

Find the loss of kinetic energy in this collision.

Solution

Using the conservation of linear momentum,

$m_A u_A + m_B u_B = m_A v_A + m_B v_B$

$4 \times 2 + 2 \times -3 = 4 \times -1 + 2v_B$

$8 - 6 = -4 + 2v_B$

$6 = 2v_B$

$v_B = 3 \, \text{m} \, \text{s}^{-1}$

Initial kinetic energy of $A = \frac{1}{2} m_A u_A{}^2 = \frac{1}{2} \times 4 \times 2^2 = 8 \, \text{J}$

Initial kinetic energy of $B = \frac{1}{2} m_B u_B{}^2 = \frac{1}{2} \times 4 \times (-3)^2 = 18 \, \text{J}$

Total initial kinetic energy $= 8 + 18 = 26 \, \text{J}$

Final kinetic energy of $A = \frac{1}{2} m_A v_A{}^2 = \frac{1}{2} \times 4 \times (-1)^2 = 2 \, \text{J}$

Final kinetic energy of $B = \frac{1}{2} m_B v_B{}^2 = \frac{1}{2} \times 4 \times 3^2 = 18 \, \text{J}$

Total final kinetic energy $= 2 + 18 = 20 \, \text{J}$

Loss of kinetic energy $= 26 - 20 = 6 \, \text{J}$

Potential energy

Potential energy is the energy a particle has as a result of its height. It is the same as the work done against gravity, from **Section 5.1,** and hence is given by mgh.

$$PE = mgh$$

It is the work done when a particle of mass m kg is raised through a vertical height of h m. You can only refer to the *change* in potential energy – it is not possible to state the actual potential energy of a particle, but you may define a vertical position to have a zero level. If a particle moves upwards, then the work done is against gravity. If the particle moves downwards, then the work done is by gravity.

> **KEY INFORMATION**
>
> The formula for the change in potential energy for a particle of mass m kg moving vertically upwards h m is PE = mgh.

> **KEY INFORMATION**
>
> You can only find the change in potential energy, not the actual potential energy.

Example 8

A flowerpot of mass 800 g falls from a shelf 240 cm above the ground. Find the loss of potential energy of the flowerpot.

Solution

The change in potential energy is given by

change in PE = mgh.

You are given that $m = 0.8$ kg and $h = 2.4$ m.

Change in PE = mgh = $0.8 \times 10 \times 2.4 = 19.2$ J.

> Change the units from g to kg and from cm to m.

When an object moves through the air or on a smooth slope (with no friction), you can often assume the rule that kinetic energy is completely converted into potential energy if the object is moving upwards (or vice versa if it is moving downwards).

> **Stop and think**
>
> What assumptions are you making when using this rule? In reality, what else will the kinetic energy be converted into? Show that if this rule is assumed to be true, then $v = \sqrt{2gh}$.

Example 9

A rocket of mass 3 kg is launched from the ground at $32 \, \text{m s}^{-1}$ and travels vertically.

Find the maximum height reached by the rocket by using:

a the equations of uniformly accelerated motion

b the change in energy.

Solution

a You could have solved this question in **Chapter 2 Kinematics of motion in a straight line**.

$u = 32 \, \text{m s}^{-1}$, $v = 0 \, \text{m s}^{-1}$, $a = -10 \, \text{m s}^{-2}$

$$v^2 = u^2 + 2as$$

$$s = \frac{v^2 - u^2}{2a}$$

$$= \frac{0^2 - 32^2}{2 \times -10} = \frac{-1024}{-20} = 51.2 \, \text{m}$$

b The change in kinetic energy is given by

$$\frac{1}{2}m(v^2 - u^2) = \frac{1}{2} \times 3(0^2 - 32^2) = -1536 \, \text{J}$$

Assume that the kinetic energy lost is all converted into a gain in potential energy. The gain in potential energy is given by mgh.

$$1536 = 3 \times 10 \times h$$

$$h = \frac{1536}{30}$$

$$= 51.2 \, \text{m}$$

Exercise 5.2A

1 **a** A particle has a mass of 8 kg and is travelling at $1.5 \, \text{m s}^{-1}$. Calculate its kinetic energy.

    b Find the kinetic energy of a car of mass 1200 kg moving at $10 \, \text{m s}^{-1}$. Give the answer in kilojoules (kJ).

    c A particle of mass 450 g has an initial velocity of $3 \, \text{m s}^{-1}$ and a final velocity of $9 \, \text{m s}^{-1}$. Find the change in kinetic energy, stating whether it is an increase or a decrease.

    d Find the change in kinetic energy when a parcel of mass 15 kg reduces in speed from $6.3 \, \text{m s}^{-1}$ to $2.7 \, \text{m s}^{-1}$, stating whether it is an increase or a decrease.

2 **a** A box of mass 4 kg is lifted 2.3 m vertically. Find the change in potential energy, stating whether it is an increase or a decrease.

    b A particle is raised vertically through 12 m. Given that the particle has a mass of 0.9 kg, find the increase in the potential energy of the particle.

    c A man of mass 80 kg climbs 4 m down a ladder. Find the decrease in the potential energy of the man.

    d A lift descends vertically through 35 m. The lift has a mass of 0.52 tonnes and the two passengers each have a mass of 75 kg. Find the change in potential energy, stating whether it is an increase or a decrease.

3 A car weighs 16 kN. The car slows down from a speed of $63 \, \text{km h}^{-1}$ to $45 \, \text{km h}^{-1}$. Find the decrease in kinetic energy as a result of this deceleration, giving the answer in kilojoules (kJ).

4 A bullet of mass 0.3 g is shot from a pistol at a speed of $400 \, \text{m s}^{-1}$. Given that the bullet decelerates at $0.2 \, \text{m s}^{-2}$ as it travels 500 m, find the loss of kinetic energy.

5 A particle of weight 45 N travels 23 m and gains 828 J of kinetic energy as it increases in speed from $19\,\mathrm{m\,s^{-1}}$ to $V\,\mathrm{m\,s^{-1}}$.

   a Find the value of V.

   b Hence find the acceleration of the particle.

5.3 Conservation of energy

A fundamental principle called the **conservation of energy** states that energy cannot be created or destroyed, only converted from one form to another. For example, when you press a light switch, electrical energy is converted to heat energy and light energy. In this section, the examples and questions focus on converting between the three types of energy you have met so far in this chapter: kinetic energy, potential energy and work done either by or against a force.

In this first section, one form of energy will be converted to another form of energy, and problems will be solved by putting them equal to each other.

> **KEY INFORMATION**
>
> The principle of conservation of energy states that energy cannot be created or destroyed, only converted from one form to another.

Example 10

A man drags a sack across a smooth horizontal patio. Initially, the sack is at rest. The man exerts a force of 30 N. After the sack has been dragged 60 cm, it is moving at $1.2\,\mathrm{m\,s^{-1}}$.

a Find the work done by the man.

b Find the mass of the sack.

Solution

a The formula for work done is WD = Fd.

   WD = $30 \times 0.6 = 18\,\mathrm{J}$ 60 cm = 0.6 m

b Because the patio is smooth, you may assume that all the work done by the man is converted into kinetic energy.

   The gain in kinetic energy is given by $\frac{1}{2}m(v^2 - u^2)$.

   $\frac{1}{2}m(v^2 - u^2) = \frac{1}{2} \times m \times (1.2^2 - 0^2) = 0.72m$

   Hence $0.72m = 18$.

   $m = \dfrac{18}{0.72} = 25\,\mathrm{kg}$

Example 11

A skateboard of mass 3.2 kg freewheels down a smooth footpath inclined at an angle of 12° to the horizontal. The initial speed of the skateboard is $2\,\mathrm{m\,s^{-1}}$. By considering the conservation of energy, find the distance the skateboard has rolled when it has a speed of $8\,\mathrm{m\,s^{-1}}$.

The term freewheel means that the skateboard has been released from rest and is just rolling down the footpath. The only force producing motion is the weight of the skateboard. Similarly, if a person is freewheeling on a bicycle down a slope, then they are not pedalling and it is just the weight that is producing the motion.

Solution

Because the footpath is smooth, you may assume that all the potential energy lost by the skateboard as it rolls down the footpath is converted into kinetic energy.

Hence loss of potential energy = gain in kinetic energy.

The loss of potential energy is given by mgh.

$mgh = 3.2 \times 10 \times h$

The vertical distance moved by the skateboard can be found by using trigonometry, as shown in **Example 5**.

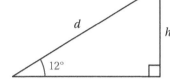

$h = d\sin 12°$

In reality, some energy will be lost as heat, sound and work done against friction, for example.

Recall that $h = d\sin\theta$.

The loss of potential energy = $32d\sin 12°$.

The gain in kinetic energy is given by $\frac{1}{2}m(v^2 - u^2)$.

$\frac{1}{2}m(v^2 - u^2) = \frac{1}{2} \times 3.2 \times (8^2 - 2^2) = 96\,\mathrm{J}$

Hence $32d\sin 12° = 96$.

$$d = \frac{96}{32\sin 12°} = 14.4\,\mathrm{m}$$

Some situations involve a combination of kinetic energy, potential energy and work done.

By the conservation of energy, the sum of the initial KE and PE will be the same as the sum of the final KE and PE and also any work done against resistive forces (noting that the change in PE is the work done against or by gravity).

Change in kinetic energy = $\frac{1}{2}m(v^2 - u^2)$

> If $v > u$, then the kinetic energy increases and the formula has a positive value.

> if $v < u$, then the kinetic energy decreases and the formula has a negative value.

Change in potential energy = mgh

> If the particle moves up, then the potential energy increases.

> If the particle moves down, then the potential energy decreases.

Although these formulae were found by assuming that the acceleration is constant in a straight line, they also apply for variable acceleration and when motion is on a curved path where the SUVAT equations do not apply but work/energy methods can be used.

Work done = *Fd*

> If work is done against a force, then the work done is positive.

> If work is done by force, then the work done is negative.

Using these rules, the conservation of energy can be written as

change in KE + change in PE + work done = 0

$$\frac{1}{2}m(v^2 - u^2) + mgh + \text{WD} = 0.$$

KEY INFORMATION

change in KE + change in PE + work done = 0

Example 12

Starting from rest, a skateboarder freewheels 15 m down a concrete slope inclined at an angle α of $\sin^{-1}\left(\frac{20}{29}\right)$. At the bottom of the slope the ground is horizontal and the skateboarder travels a further 21 m until she gets to another concrete slope inclined at an angle β of $\sin^{-1}\frac{5}{13}$, as shown in the diagram.

The skateboarder experiences a constant resistive force of 14 N throughout. The total mass of the girl and her skateboard is 75.4 kg. By considering energy changes, find the distance the girl skates up the second slope before she comes to rest.

Solution

Since the initial and final velocities of the skateboarder are $0\,\text{m s}^{-1}$, the initial KE and final KE are 0 J, and hence the change in KE = $0 - 0 = 0$ J.

Relative to the horizontal ground, the initial potential energy of the girl = $mgh = 75.4 \times 10 \times 15 \sin \alpha$.

$75.4 \times 10 \times 15 \sin \alpha = 75.4 \times 10 \times 15 \times \frac{20}{29} = 7800\,\text{J}$

Relative to the horizontal ground, the final potential energy of the girl = $mgh = 75.4 \times 10 \times d \sin \beta$.

$75.4 \times 10 \times d \sin \beta = 754d \times \frac{5}{13} = 290d\,\text{J}$

Note that the PE must have decreased, since the final height of the skateboarder must be lower than her initial height if there is no change in KE and work has been done against the resistive force (otherwise WD would be negative).

Note that it does not matter that the skateboarder has kinetic energy during the motion.

Change in PE = (290d – 7800) J.

The work done against the resistive force,
WD = Fd = 14(15 + 21 + d).

Substitute into the equation for the conservation of energy.

Change in KE + change in PE + work done = 0

0 + (290d – 7800) + 14(15 + 21 + d) = 0

290d – 7800 + 210 + 294 + 14d = 0

290d + 210 + 294 + 14d = 7800

304d = 7296

$$d = \frac{7296}{304} = 24\,\text{m}$$

Stop and think Comment on the assumption that the resistive force is constant. How could you test or refine this model?

Exercise 5.3A

1 A football of mass 0.4 kg is kicked vertically upwards from the ground. The initial speed of the ball is 18 m s^{-1}. Assume that there are no resistive forces.

   a Find the initial kinetic energy of the football.

   b Find the potential energy gained by the football when it reaches its maximum height.

   c Find the height of the ball above the ground when it stops instantaneously.

2 Starting from rest, a boy of mass 30 kg slides 5.4 m down a slide at a playground. The slide is rough, with a coefficient of friction of 0.15 and an angle of inclination, θ, of $\tan^{-1}\left(\frac{3}{4}\right)$.

   a State the initial kinetic energy of the boy.

   b Find the change in the potential energy of the boy.

   c Find the normal reaction force exerted by the slide on the boy.

   d Find the frictional force.

   e Find the work done against friction.

   f Find the speed of the boy at the bottom of the slide.

3 A brick of mass 2.8 kg falls from the top of a block of flats. When it passes the tenth floor it has a speed of 11 m s^{-1}. When it passes the third floor it has a speed of 25 m s^{-1}. The floors are evenly spaced out.

   a Find the change in the brick's kinetic energy between the tenth and third floors.

   b By considering the loss of potential energy, find the distance between the fifth and seventh floors.

4 A bowling ball has a mass of 1.6 kg. It rolls along a rough floor. Its initial velocity is 8 m s⁻¹ and it rolls 48 m before coming to rest.

   a Find the initial kinetic energy of the bowling ball.

   b Find the work done against friction.

   c Find the coefficient of friction between the ball and the floor.

5 A bullet of mass 2.5 g is shot at 300 m s⁻¹. It immediately hits a wall and experiences a resistive force of 2.25 kN. Find how far into the wall the bullet penetrates.

6 A pebble of mass 40 g is dropped from the top of a tower of height 60 m.

   a By considering the energy of the pebble, find the speed of the pebble as it hits the ground.

   b What assumption have you made in answering **part a**?

   c Show that you get the same answer as in **part a** if you use the equation $v^2 = u^2 + 2as$.

7 A particle of mass 0.7 kg is projected at 7 m s⁻¹ up a rough plane inclined at 7° to the horizontal. The particle moves 18 m up the plane before coming to rest. Find the coefficient of friction between the particle and the plane.

8 A boy and his bicycle have a total weight of 66 kg. Starting from rest, the boy freewheels d m down a road inclined at an angle of θ, where $\theta = \sin^{-1}\left(\frac{1}{5}\right)$. At the bottom of the slope the road becomes horizontal and the boy travels a further d m until he gets to a second slope at the same angle as the first, as shown in the diagram, and he stops 6 m up the slope.

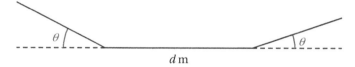

d m

The boy experiences a constant resistive force of 12 N throughout.

   a By considering energy changes, find the value of d.

   b Find the speed at which the boy is travelling halfway along the horizontal section of road.

9 Starting from point X, a particle P, of mass 2 kg, is projected from rest down a rough curved surface, XYZ, as shown in the diagram, where $XY = 20$ m and $YZ = 10$ m.

The vertical height of the surface between X and Y is 13 m and between Y and Z is H m. The particle experiences a constant resistive force of 4 N. The particle comes to rest at Z.

Use the conservation of energy to find:

   a the speed of P at point Y

   b the value of H.

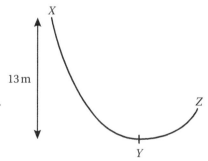

13 m

10 A cart of mass 8.5 kg is projected at 14 m s⁻¹ up a rough ramp inclined at an angle θ to the horizontal, where $\sin \theta = \frac{3}{5}$. The ramp is 12 m wide, as shown in the diagram.

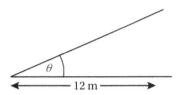

The cart stops instantaneously at the top of the ramp, By considering the conservation of energy, find the frictional force on the cart.

11 A particle is projected up a rough plane inclined at an angle β to the horizontal, where $\sin \beta = \frac{3}{5}$. The height of the plane is 1.6 m. The coefficient of friction between the particle and the plane is $\frac{3}{16}$.

Given that the final velocity of the particle is given by v m s⁻¹ and the initial velocity of the particle by $(2v + 1)$ m s⁻¹, use the conservation of energy to find the value of v.

12 A builder standing on the roof of a house dislodges a slate of mass 0.6 kg, which then slides down the roof and falls onto the ground and breaks. The roof is inclined at an angle of $\tan^{-1}\left(\frac{\sqrt{2}}{4} \right)$ to the horizontal and is modelled as a rough surface with a coefficient of friction of $\frac{\sqrt{2}}{2}$ between the roof and the slate. The roof is 3 m tall and the height of the whole house is 12 m. The speed of the slate when it hits the ground is 14 m s⁻¹.

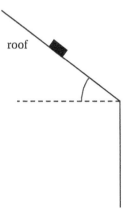

roof

Find the distance the slate slides down the roof.

13 A boy is skateboarding in a half pipe, as shown in the diagram.

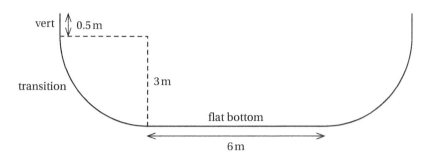

The combined mass of the boy and his skateboard is 48 kg. From a standing start, he sets off from the edge of the deck and freewheels down the vert (vertical), into the transition and flat bottom and up the other transition against a constant resistive force. The vert is 0.5 m, the transition is a quadrant of a circle with a radius of 3 m and the flat bottom is 6 m wide. The boy comes to rest at the top of the second transition.

a Find the magnitude of the resistive force.

b If the resistive force were 10 N instead, find the distance that the boy would travel up the second vert.

14 A ball is released from point *A* at the top of a wooden track, *AB*, inclined at an angle of 20° to the horizontal, as shown in the diagram.

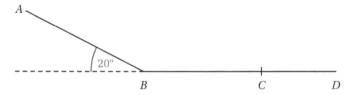

BD is also a wooden track, but horizontal. The coefficient of friction between the ball and each track is 0.3. The ball comes to rest at point *C*. *BC* = 6 m.

a Find the speed of the ball at *B*.

b Find the length of *AB*.

c Find the time taken for the ball to move from *A* to *C*.

Mathematics in life and work: Group discussion

You work in research and development for a top international motor racing team that competes in races around the world. You are developing a racing car, with the requirement that it can reach up to 300 km h⁻¹ on a straight section of track.

1 Write the maximum speed, using SI units.

2 What factors influence the magnitudes of the forces acting upon the car?

3 Given that the car enters a straight at 120 km h⁻¹ and that the car experiences a resistance to motion of 1200 N, estimate the work that needs to be done by the engine of the car to get to 300 km h⁻¹ within 100 m.

4 How would your answer be affected by the slope of the straight?

5.4 Power

In **Section 5.1**, work done was related to the amount of effort required to push a trolley and the longer the distance the more work was required. If the trolley is pushed the same distance but in a shorter time, the same amount of work is required but because the trolley is being pushed faster it requires more **power**. Power is the rate of doing work and is given by the formula

average power $= \dfrac{\text{WD}}{t}$

where WD is the work done (in joules, J) and t is the time taken (in seconds, s). The units for power are **watts**, W. Since power is a rate of change of energy, the units can also be written as J s^{-1}, N m s^{-1} and $\text{kg m}^2 \text{s}^{-3}$.

From this equation, you can derive another equation for power, which is more useful when considering the driving force of a vehicle.

Because work done is given by force × distance,

$P = \dfrac{\text{WD}}{t} = \dfrac{Fd}{t}$

At a particular instant, the velocity of a vehicle is given by the formula velocity $= \dfrac{d}{t}$.

Hence, $P = Fv$, where F is the force and v is the velocity.

Example 13

A girl and a cart have a combined mass of mass 80 kg. The cart is moving along a rough horizontal floor. The coefficient of friction between the cart and the floor is $\frac{2}{5}$. The engine of the cart is working at a rate of 2.4 kW.

Find:

a the acceleration of the cart when it is moving at $5\,\text{m s}^{-1}$

b the maximum velocity of the cart

c the work done when the cart is moving at its maximum velocity for 8 s.

Solution

There are four forces acting upon the cart: friction, reaction, weight and driving force.

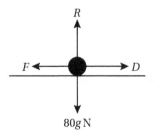

a Resolve in the direction of motion.

$R(\rightarrow)$

$D - F = ma$

The driving force is given by $D = \dfrac{P}{v}$.

The frictional force can be found by finding an expression for R and then using $F = \mu R$.

So $\dfrac{P}{v} - \mu R = 80a$

Resolve vertically.

$R(\uparrow)$

$R - 80g = 0$

$R = 80g = 80 \times 10 = 800\,\text{N}$

$F = \mu R = \dfrac{2}{5} \times 800 = 320\,\text{N}$

Hence $\dfrac{2400}{v} - 320 = 80a$.

$\dfrac{2400}{5} - 320 = 80a$

$80a = 160$

$a = 2\,\text{m s}^{-2}$

b When the cart is travelling at its maximum velocity, the acceleration of the cart is zero.

$\dfrac{P}{v} - \mu R = ma$

$\dfrac{2400}{v} - 320 = 0$

$v = \dfrac{2400}{320} = 7.5\,\text{m s}^{-1}$

c Using the formula

$P = \dfrac{\text{WD}}{t}$

$\text{WD} = Pt = 2400 \times 8 = 19\,200\,\text{W}$.

> If $P = Fv$, then $F = \dfrac{P}{v}$.

> **KEY INFORMATION**
> When the velocity is a maximum, the acceleration is zero.

Example 14

A car of mass 1200 kg moves up a road inclined at an angle of 9° to the horizontal. The coefficient of friction between the car and the road is 0.06. Find the power developed when the car is moving at $15\,\mathrm{m\,s^{-1}}$ at an acceleration of $0.3\,\mathrm{m\,s^{-2}}$.

Solution

The same four forces are acting upon the car: friction, reaction, weight and driving force.

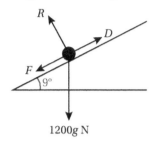

a Resolve in the direction of motion.

R(↗)

$D - F - mg\sin 9° = ma$

Hence $F = D - mg\sin 9° - ma$.

Resolve perpendicular to the surface.

R(↖)

$R - mg\cos 9° = 0$

Hence $R = mg\cos 9°$.

Substitute into $F = \mu R$.

$D - mg\sin 9° - ma = \mu mg\cos 9°$

$\dfrac{P}{15} - 1200 \times 10\sin 9° - 1200 \times 0.3 = 0.06 \times 1200 \times 10\cos 9°$

$P = 15(12\,000\sin 9° + 360 + 720\cos 9°) = 44.2\,\mathrm{kW}$

Exercise 5.4A

1 a Find the power developed in a car travelling at $14\,\mathrm{m\,s^{-1}}$, given that its engine is producing a driving force of 1500 N.

b A child is pushing a toy pushchair across the floor with a force of 10 N. The pushchair is moving at $0.2\,\mathrm{m\,s^{-1}}$. Find the power developed by the child.

c The engine of a truck is working at a rate of 7200 W. Given that the truck is travelling at $54\,\mathrm{km\,h^{-1}}$, find the driving force produced by the engine.

d The engine of a van is producing a driving force of 1.8 kN and working at a rate of 9.9 kW. Find the velocity of the van.

e The work done by a crane to lift a steel girder is 20 kJ. If it takes 10 s to lift the girder, find the power developed.

f An engine works at a rate of 6.1 kW for 29 s. Find the work done by the engine.

2 A motorcycle of mass 250 kg is being ridden across a horizontal dirt track by a woman with a mass of 55 kg. The coefficient of friction between the motorcycle and the track is $\frac{2}{5}$.

a Find the normal reaction force exerted on the motorcycle by the dirt track.

b Find the frictional force.

The woman rides 100 m in 5 s at the motorcycle's maximum velocity.

c Find the work done by the engine.

d Find the power developed by the motorcycle, giving the answer in kilowatts (kW).

e Find the acceleration of the motorcycle when it is moving at 12.5 m s^{-1} with the same power developed by the engine as in **part d**.

3 A car of mass 1200 kg is moving along a rough horizontal road. The engine is working at a rate of 7.2 kW. The coefficient of friction between the car and the road is $\frac{1}{24}$.

Find:

a the velocity when the car is accelerating at 1.25 m s^{-2}

b the velocity when the car is accelerating at 0.25 m s^{-2}

c the maximum velocity of the car.

4 A car of mass 1.3 tonnes is driven up a rough road inclined at 6° to the horizontal. The engine of the car works at a rate of 14.8 kW. The coefficient of friction between the car and the road is 0.03.

a Find the maximum velocity of the car.

b Find the acceleration of the car when the velocity is 6 m s^{-1}, assuming that the engine works at the same rate.

5 A minibus of mass 3200 kg is driven up a rough slope inclined at an angle of $\sin^{-1}\left(\frac{3}{16}\right)$ to the horizontal. It is subject to a constant resistance of 800 N. Given that the maximum velocity as it is driven up the slope is 1.8 m s^{-1},

a find the power developed by the engine of the minibus.

The slope then becomes horizontal. Given that the power developed by the engine and the resistance are the same,

b find the initial acceleration of the minibus.

6 A van is driven along a rough horizontal road at a maximum speed of $20\,\text{m s}^{-1}$. The engine of the van is working at $8.4\,\text{kW}$. The coefficient of friction between the van and the road is $\frac{1}{25}$.

  a Find the mass of the van.

  b Find the work done by the van when it is driven for 30 seconds.

The road is now modelled as a rough slope inclined at 3° to the horizontal. The power developed by the engine and the coefficient of friction are assumed to be the same.

  c Find the new maximum speed of the van.

  d Find the acceleration of the van when it is moving up the hill at $5\,\text{m s}^{-1}$.

7 A car of mass $800\,\text{kg}$ tows another car of mass $1000\,\text{kg}$ up a hill. The engine of the first car is working at a rate of $9.4\,\text{kW}$. The two vehicles are moving up the hill at $5\,\text{m s}^{-1}$ and accelerating at $0.4\,\text{m s}^{-2}$. The cars are modelled as particles, each subject to a constant resistive force of $400\,\text{N}$, and the hill is modelled as a plane inclined at an angle of θ to the horizontal.

  a Show that $\sin\theta = 0.02$.

  b Hence find the maximum possible velocity, given that the power developed remains the same.

The vehicles become uncoupled when travelling at maximum velocity.

  c Find the acceleration of the first car immediately after the uncoupling.

8 Fen is riding her motorbike up a rough slope inclined at an angle of 4° to the horizontal. The combined mass of Fen and her motorbike is $180\,\text{kg}$. The power developed by the motorbike is $6.8\,\text{kW}$. When the motorbike is accelerating at $0.2\,\text{m s}^{-2}$, it has a velocity of $18\,\text{ms}^{-1}$. The coefficient of friction between the slope and the motorbike is given by μ.

  a Find the value of μ.

When Fen is riding at maximum velocity up the same slope, she goes into neutral (so that the engine is no longer working).

  b Use conservation of energy to find the further distance Fen travels up the slope before she stops instantaneously.

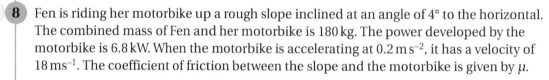

9 A van of mass 1.6 tonnes is travelling along a horizontal road at $V\,\text{m s}^{-1}$. The frictional force is directly proportional to the velocity of the van. The power generated by the engine is $28.8\,\text{kW}$. When $V = 12\,\text{m s}^{-1}$, the car accelerates at $1.26\,\text{m s}^{-2}$.

  a Find a formula for the frictional force in terms of V.

  b Find the value of V when the van is accelerating at $0.5\,\text{m s}^{-2}$.

10 A car of mass $1200\,\text{kg}$ travels along a straight road inclined at 3° to the horizontal. The car experiences a frictional force of $40V\,\text{N}$, where V is the velocity of the car in m s^{-1}.

Given that the power developed by the engine of the car in this situation is $75\,\text{kW}$,

  a find the maximum velocity of the car.

The inclination of the road is now θ, where $\sin\theta = \frac{1}{20}$. The power developed has reduced by $30\,\text{kW}$ and the resistive force has fallen to $30V$.

  b Find the new maximum velocity of the car.

Mathematics in life and work: Group discussion

In order for a racing car to reach high speeds, the engine needs to work at a high rate. In your position in research and development for a top international motor racing team, you are investigating the requirements that your car needs to satisfy in order to make it as competitive as possible.

1 Find the power developed by the engine of a racing car that has a maximum speed of $300 \, \text{km h}^{-1}$ against a resistance of $1200 \, \text{N}$ on a horizontal track.

2 Find the power developed if the slope is inclined at:

   a 2°   **b** 4°   **c** 6°.

3 What considerations would you need to make as the power developed increases?

4 If you needed to keep the power developed to a maximum of $25 \, \text{kW}$, what would be the maximum resistance you could allow on the horizontal?

5 What steps could you take to reduce the resistance between the car and the track?

SUMMARY

> The work done by or against a force is given by the formula WD = Fd, where F is the force (in newtons) and d is the distance (in m). When the force acts at an angle θ, WD = $Fd\cos\theta$.

> Kinetic energy is the energy a particle has as a result of its motion. KE = $\frac{1}{2}mv^2$ and the change in kinetic energy = $\frac{1}{2}m(v^2 - u^2)$, where m is the mass (in kg) and u and v are velocities (in m s^{-1}).

> Potential energy is the energy a particle has as a result of its height.
> PE = mgh, where h is the height (in m). Only the change in potential energy can be measured, not the actual potential energy.

> The unit for energy is the joule (J).

> The principle of conservation of energy states that energy cannot be created or destroyed, only converted from one form to another. Change in KE + change in PE + work done = 0.

> Average power = $\dfrac{\text{WD}}{t}$, where WD is the work done (in J) and t is the time taken (in s).

> Power = Fv, where F is the force (in N) and v is the velocity (in m s^{-1}).

> The unit for power is the watt (W).

EXAM-STYLE QUESTIONS

1 A coin is dropped into a well. The mass of the coin is 3 g and the height of the well is 26 m.

   a Find the loss of potential energy of the coin.

   b Find the speed of the coin when it gets to the bottom of the well.

2 A suitcase of mass 1.6 kg is projected 5 m along an airport floor until it comes to rest. Initially, the suitcase has 7.2 J of kinetic energy. The suitcase is modelled as a particle and the floor is modelled as rough and horizontal.

   a Find the speed at which the suitcase is projected.

   b State the work done against the resistive force.

   c Find the magnitude of the resistive force.

3 A remote-controlled car is driven across a tiled floor. The mass of the car is 1.3 kg. The car experiences a constant resistive force of 3 N. The battery in the car works at rate of 7.5 W.

   a Find the driving force when the car is driven at its maximum velocity.

   b Find the maximum velocity of the remote-controlled car.

   c Find the acceleration of the remote-controlled car when it is travelling at 1.2 m s^{-1}.

4 A ball is thrown vertically upwards, with a speed of 24 m s^{-1}, from a height of 6 m above the ground. Assuming that all of the kinetic energy is converted into potential energy, find the maximum height reached by the ball above the ground.

5 A bucket is accidentally dropped from the top of a building. When the bucket is h m above the ground it is falling at $22\,\text{m s}^{-1}$. When the bucket hits the ground it is falling at $28\,\text{m s}^{-1}$. The bucket gains 360 J of energy as it falls the final h m.

    a Find the mass of the bucket.

    b Find the value of h.

    c Find the time taken for the bucket to hit the ground.

    d Find the height of the building.

6 A car that has run out of petrol is being towed by a breakdown truck along a straight horizontal road. The truck has mass of 1600 kg and the car has mass of 900 kg. The truck is connected to the car by a horizontal rope that is modelled as light and inextensible. The truck's engine provides a constant driving force of 2500 N. The resistances to motion of the truck and the car are modelled as constant and of magnitude 500 N and 300 N, respectively.

    Find:

    a the acceleration of the truck

    b the tension in the rope.

    c Explain how you used the modelling assumption that the rope is inextensible.

    When the breakdown truck is moving at $25\,\text{m s}^{-1}$, the rope breaks. The magnitude of the resistance to the motion of the car is assumed not to have changed.

    d Use conservation of energy to find the distance travelled by the car after the rope breaks.

7 A golf buggy of mass 225 kg is driven along a rough horizontal lawn at maximum speed. The engine of the buggy is working at 2.16 kW. The coefficient of friction between the golf buggy and the lawn is 0.16.

    a Find the maximum speed of the golf buggy.

    The golf buggy is now driven up a grass verge in the rough, inclined at an angle of 5° to the horizontal. The engine works at the same rate but the coefficient of friction is now 0.2.

    b Find the maximum speed of the golf buggy on the grass verge.

8 A ball of mass 250 g is released from rest at the top of a rough plane that is inclined at $\cos^{-1} 0.8$ to the horizontal. The coefficient of friction between the particle and the plane is $\frac{3}{7}$. The plane is of length 2.5 m. Find, at the point when the ball reaches the bottom of the plane:

    a the potential energy lost by the ball

    b the work done against friction by the ball

    c the speed of the ball at the bottom of the plane.

9 A toy robot of mass 2.5 kg is projected up a rough slope inclined at $\sin^{-1}\left(\frac{57}{1625}\right)$ to the horizontal. The robot comes to instantaneous rest after 6.5 m. The coefficient of friction between the robot and the slope is $\frac{5}{232}$.

Find:

a the work done against gravity

b the total work done.

The robot then slides back down the slope.

c Find the speed of the robot when it returns to its starting point.

10 Two particles, A and B, approach each other at speeds of $2\,\mathrm{m\,s^{-1}}$ and $4\,\mathrm{m\,s^{-1}}$, respectively. After the particles collide, they are travelling in different directions, A at $2.5\,\mathrm{m\,s^{-1}}$ and B at $3.5\,\mathrm{m\,s^{-1}}$. The mass of particle B is 3 kg.

a Find the mass of particle A.

b Show that there is no loss of kinetic energy as a result of the collision.

11 Two particles, P and Q, of masses 4 kg and 6 kg, respectively, are connected by means of a light inextensible string which passes over a smooth fixed pulley. Q is suspended 2 m above the ground. P is at rest on a rough slope 3 m from the pulley. The angle between the two ends of the string is 60°, as shown in the diagram.

The system is released. Q falls to the ground. As P is pulled up the slope it is subject to a constant resistive force of 30 N.

a Use conservation of energy to find the velocity of Q when it hits the ground.

b Find the total distance travelled by P before it comes to instantaneous rest.

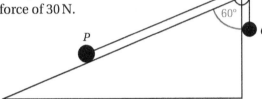

12 A lorry pulls a cement mixer along a horizontal path. Due to the relative sizes of the vehicles, the coupling is inclined at 9° to the horizontal. The lorry and cement mixer have masses of 1600 kg and 800 kg, respectively. The coefficient of friction between each vehicle and the path is $\frac{1}{16}$. The lorry sets off from a stationary position with a constant driving force of 3.3 kN.

a Find the initial acceleration of the lorry and cement mixer.

b Find the tension in the coupling.

After travelling 24 m, the coupling becomes disconnected.

c Find the power developed by the engine when the vehicles become uncoupled.

d Assuming that the power developed remains the same after the vehicles are uncoupled, find the maximum velocity of the lorry.

e Find the distance travelled by the cement mixer before it rolls to a stop.

13 A stuntman on a bicycle starts his act from point X on the diagram, above a steeply curved concrete slope.

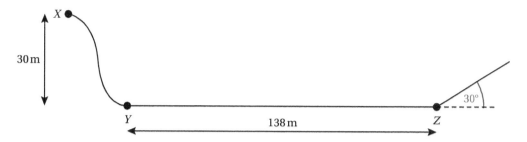

From a standing start, he freewheels down the slope to point Y. At point Y, the concrete becomes horizontal for 138 m to point Z. At point Z, there is a 30° incline. The combined mass of the stuntman and his bicycle is 200 kg. The bicycle is subject to a constant resistance of 232 N. The distance XY is 50 m and X is 30 m above the ground.

a Find the speed at which the stuntman is travelling at Y.

b Find the speed at which the stuntman is travelling at Z.

c Find the distance the stuntman travels up the slope before coming to rest.

14 A tractor of weight 25 000 N pulls a combine harvester up a rough slope inclined at 10°. The coefficient of friction between the slope and each vehicle is 0.125. The engine of the tractor is working at a rate of 48 kW. The maximum speed of the vehicles is 4.2 m s^{-1}.

a Find the weight of the combine harvester.

b Find the tension in the coupling.

Suddenly the vehicles become uncoupled.

c Find the time it takes for the combine harvester to come to instantaneous rest.

15 Angelene skateboards along a track. The track starts and finishes with hard slopes AB and CD, each inclined at θ to the horizontal, where $\sin \theta = \frac{3}{5}$, and 10 m long, as shown in the diagram.

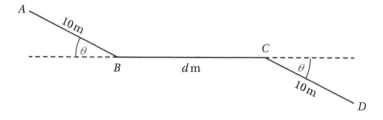

Between the two slopes there is a horizontal section BC of length d m. The coefficient of friction between the skateboard and the track is $\frac{1}{10}$ for all three sections. Angelene starts from rest at A and freewheels all the way. She has just enough motion at C to continue down the final section.

a Find the value of d.

Given that Angelene has 3.12 kJ of kinetic energy at D:

b find the combined mass of Angelene and her skateboard

c find the speed at which Angelene is moving when she gets to D.

16 Two particles, P and Q, of masses 9 kg and 7 kg, respectively, are connected by a light inextensible string. The particles are suspended vertically with P above Q, and Q is 5 m above the ground. A force of 200 N is applied vertically upwards on P such that P is raised 8 m.

Find:

a the work done by the 200 N force

b the work done against gravity

c the speed of the particles when they have been raised 8 m.

When the string has been raised 8 m, the string snaps.

d Find the speed of Q when it hits the ground.

17 A car of mass 1800 kg is moving along a horizontal road. The power developed by the engine of the car is 60.2 kW. The resistive force experienced by the car as it moves along the road is given by $(28v + 20)$ N, where v ms^{-1} is the velocity of the car. When the car is accelerating at 0.4 ms^{-2}, the velocity is given by V ms^{-1}.

a Show that $7V^2 + 185V = 15050$.

b Hence find the value of V.

c Show that the maximum velocity of the car is about 11 ms^{-1} faster than V.

18

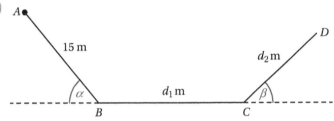

A particle of weight 6.8 kg moves 15 m down a plane from A to B, along a horizontal path d_1 m to C, then d_2 m up a second plane until it comes to instantaneous rest at D, as shown in the diagram. The particle is subject to a constant frictional force of F N throughout. The first and second planes are inclined at respective angles α and β to the horizontal, where $\sin \beta = \dfrac{13}{85}$. The particle has an initial gravitational potential energy of 480 J relative to line BC.

The total distance travelled by the particle is 65 m and d_1 and d_2 are in the ratio $3 : 1$.

a Find the value of $\sin \alpha$, giving the answer as a simplified fraction.

b Find the gravitational potential energy of the particle at D.

c Find the value of F, correct to 1 decimal place.

d Find the velocity of the particle at C.

19 The heights of the front and back of a shed are h m and $\frac{3}{2}h$ m, respectively, as shown in the diagram. A tile slides h m down the roof before falling vertically to the ground. The roof is rough and the coefficient of friction between the roof and the tile is given by $\frac{\sqrt{3}}{3}$. Use conservation of energy to show that the speed of the tile when it hits the ground is given by $\sqrt{2gh}$.

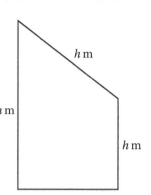

h m

$\frac{3}{2}h$ m

h m

Mathematics in life and work

You work for a top international motor racing team and you are preparing your car for an important race. One part of the track consists of a 140 m downwards slope of height 20 m, which is curved, followed by a straight slope inclined at 3° to the horizontal, as shown in the diagram.

3°

The mass of the car is 720 kg, including the driver and fuel. The car will experience a constant resistive force of 2400 N on the curved section. The coefficient of friction between the car and the track on the straight slope is 0.2.

At the start of the curved slope, the car will be moving at 288 km h⁻¹. No driving force will be applied whilst the car is travelling down the slope.

1 Find the speed of the car at the bottom of the curved slope.

When the car begins to move up the straight slope, its initial acceleration is 0.4 m s⁻².

2 Find the power developed by the engine at the beginning of the straight slope.

The engine works at a constant rate as it moves up the slope.

3 Find the maximum speed of the car on the slope.

SUMMARY REVIEW

Practise the key concepts and apply the skills and knowledge that you have learned in the book with these carefully selected past paper questions supplemented with exam-style questions and extension questions written by the authors.

Warm-up Questions	A Level Questions	Extension Questions
Three Cambridge IGCSE® past paper questions based on prerequisite skills and concepts that are relevant to the main content of this book.	Selected past paper exam questions and exam-style questions on the topics covered in this syllabus component.	Extension questions that give you the opportunity to challenge yourself and prepare you for more advanced study.

Warm-up questions

Reproduced by permission of Cambridge Assessment International Education

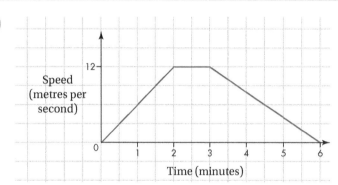

A tram leaves a station and accelerates for 2 **minutes** until it reaches a speed of 12 metres per second.

It continues at this speed for 1 minute.

It then decelerates for 3 minutes until it stops at the next station.

The diagram shows the speed–time graph for this journey.

Calculate the distance, in metres, between the two stations.

[3]

Cambridge IGCSE Mathematics 0580 Paper 21 Q10 June 2015

2

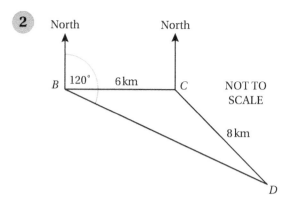

A helicopter flies from its base *B* to deliver supplies to two oil rigs at *C* and *D*.

C is 6 km due east of *B* and the distance from *C* to *D* is 8 km.

D is on a bearing of 120° from *B*.

Find the bearing of *D* from *C*.   [5]

Cambridge IGCSE Mathematics 0580 Paper 21 Q16 Nov 2014

3

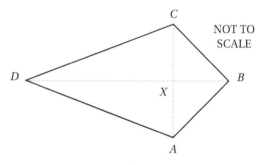

ABCD is a kite.

The diagonals *AC* and *BD* intersect at *X*.

AC = 12 cm, *BD* = 20 cm and *DX*:*XB* = 3:2.

  i Calculate angle *ABC*.   [3]

  ii Calculate the area of the kite.   [2]

Cambridge IGCSE Mathematics 0580 Paper 21 Q21 Nov 2013

A Level questions

1 A train is moving at constant speed V m s^{-1} along a horizontal straight track. Given that the power of the train's engine is 1330 kW and the total resistance to the train's motion is 28 kN, find the value of V.   [3]

Cambridge International AS & A Level Mathematics 9709 Paper 41 Q1 June 2014

2

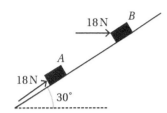

Small blocks A and B are held at rest on a smooth plane inclined at 30° to the horizontal. Each is held in equilibrium by a force of magnitude 18 N. The force on A acts upwards parallel to a line of greatest slope of the plane, and the force on B acts horizontally in the vertical plane containing a line of greatest slope (see diagram). Find the weight of A and the weight of B. [4]

Cambridge International AS & A Level Mathematics 9709 Paper 41 Q2 Nov 2014

3 A block of weight 6.1 N slides down a slope inclined at $\tan^{-1}\left(\dfrac{11}{60}\right)$ to the horizontal. The coefficient of friction between the block and the slope is $\dfrac{1}{4}$. The block passes through a point A with speed 2 m s^{-1}. Find how far the block moves from A before it comes to rest. [5]

Cambridge International AS & A Level Mathematics 9709 Paper 41 Q3 June 2015

4 A lorry of mass 14 000 kg moves along a road starting from rest at a point O. It reaches a point A, and then continues to a point B which it reaches with a speed of 24 m s^{-1}. The part OA of the road is straight and horizontal and has length 400 m. The part AB of the road is straight and is inclined downwards at an angle of $\theta°$ to the horizontal and has length 300 m.

i For the motion from O to B, find the gain in kinetic energy of the lorry and express its loss in potential energy in terms of θ. [3]

The resistance to the motion of the lorry is 4800 N and the work done by the driving force of the lorry from O to B is 5000 kJ.

ii Find the value of θ. [3]

Cambridge International AS & A Level Mathematics 9709 Paper 41 Q4 June 2015

5 A particle P moves in a straight line, starting from a point O. The velocity of P, measured in m s^{-1}, at time t s after leaving O is given by

$$v = 0.6t - 0.03t^2.$$

i Verify that, when $t = 5$, the particle is 6.25 m from O. Find the acceleration of the particle at this time. [4]

ii Find the values of t at which the particle is travelling at half of its maximum velocity. [6]

Cambridge International AS & A Level Mathematics 9709 Paper 41 Q6 Nov 2015

6 Particles *A* of mass 0.25 kg and *B* of mass 0.75 kg are attached to opposite ends of a light inextensible string which passes over a fixed smooth pulley. The system is held at rest with the string taut and its straight parts vertical. Both particles are at a height of *h* m above the floor (see Fig. 1). The system is released from rest, and 0.6 s later, when both particles are in motion, the string breaks. The particle *A* does not reach the pulley in the subsequent motion.

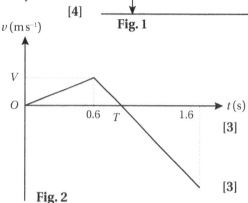

Fig. 1

i Find the acceleration of *A* and the distance travelled by *A* before the string breaks. **[4]**

The velocity–time graph shown in Fig. 2 is for the motion of particle *A* until it hits the floor. The velocity of *A* when the string breaks is *V* m s^{-1} and *T* s is the time taken for *A* to reach its greatest height.

ii Find the value of *V* and the value of *T*. **[3]**

iii Find the distance travelled by *A* upwards and the distance travelled by *A* downwards and hence find *h*. **[3]**

Fig. 2

Cambridge International AS & A Level Mathematics 9709 Paper 41 Q6 June 2014

7

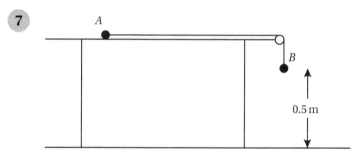

Particles *A* and *B*, of masses 0.3 kg and 0.7 kg respectively, are attached to the ends of a light inextensible string. Particle *A* is held at rest on a rough horizontal table with the string passing over a smooth pulley fixed at the edge of the table. The coefficient of friction between *A* and the table is 0.2. Particle *B* hangs vertically below the pulley at a height of 0.5 m above the floor (see diagram). The system is released from rest and 0.25 s later the string breaks. *A* does not reach the pulley in the subsequent motion. Find

i the speed of *B* immediately before it hits the floor, **[9]**

ii the total distance travelled by *A*. **[3]**

Cambridge International AS & A Level Mathematics 9709 Paper 41 Q7 June 2015

8 A weightlifter performs an exercise in which he raises a mass of 200 kg from rest vertically through a distance of 0.7 m and holds it at that height.

i Find the work done by the weightlifter. **[2]**

ii Given that the time taken to raise the mass is 1.2 s, find the average power developed by the weightlifter. **[2]**

Cambridge International AS & A Level Mathematics 9709 Paper 41 Q1 Nov 2015

9 A lorry of mass 24 000 kg is travelling up a hill which is inclined at 3° to the horizontal. The power developed by the lorry's engine is constant, and there is a constant resistance to motion of 3200 N.

    i When the speed of the lorry is 25 m s^{-1}, its acceleration is 0.2 m s^{-2}. Find the power developed by the lorry's engine. [4]

    ii Find the steady speed at which the lorry moves up the hill if the power is 500 kW and the resistance remains 3200 N. [2]

Cambridge International AS & A Level Mathematics 9709 Paper 41 Q3 Nov 2015

10

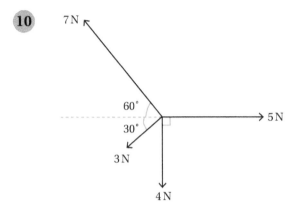

Four coplanar forces act at a point. The magnitudes of the forces are 5 N, 4 N, 3 N and 7 N, and the direction in which the forces act are shown in the diagram. Find the magnitude and direction of the resultant of the four forces. [6]

Cambridge International AS & A Level Mathematics 9709 Paper 41 Q3 June 2014

11

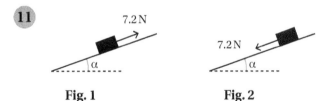

Fig. 1   **Fig. 2**

A block of weight 7.5 N is at rest on a plane which is inclined to the horizontal at angle α, where $\tan\alpha = \dfrac{7}{24}$. The coefficient of friction between the block and the plane is μ. A force of magnitude 7.2 N acting parallel to a line of greatest slope is applied to the block. When the force acts up the plane (see Fig. 1) the block remains at rest.

    i Show that $\mu \geqslant \dfrac{17}{24}$. [4]

When the force acts down the plane (see Fig. 2) the block slides downwards.

    ii Show that $\mu < \dfrac{31}{24}$. [2]

Cambridge International AS & A Level Mathematics 9709 Paper 41 Q3 Nov 2014

12 A particle of mass 3 kg falls from rest at a point 5 m above the surface of a liquid which is in a container. There is no instantaneous change in speed of the particle as it enters the liquid. The depth of the liquid in the container is 4 m. The downward acceleration of the particle while it is moving in the liquid is 5.5 m s^{-2}.

i Find the resistance to motion of the particle while it is moving in the liquid. **[2]**

ii Sketch the velocity–time graph for the motion of the particle, from the time it starts to move until the time it reaches the bottom of the container. Show on your sketch the velocity and the time when the particle enters the liquid, and when the particle reaches the bottom of the container. **[7]**

Cambridge International AS & A Level Mathematics 9709 Paper 41 Q6 Nov 2014

13

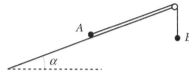

A light inextensible string has a particle A of mass 0.26 kg attached to one end and a particle B of mass 0.54 kg attached to the other end. The particle A is held at rest on a rough plane inclined at angle α to the horizontal, where $\sin \alpha = \frac{5}{13}$. The string is taut and parallel to a line of greatest slope of the plane. The string passes over a small smooth pulley at the top of the plane. Particle B hangs at rest vertically below the pulley (see diagram). The coefficient of friction between A and the plane is 0.2. Particle A is released and the particles start to move.

i Find the magnitude of the acceleration of the particles and the tension in the string. **[6]**

Particle A reaches the pulley 0.4 s after starting to move.

ii Find the distance moved by each of the particles. **[2]**

Cambridge International AS & A Level Mathematics 9709 Paper 41 Q5 June 2013

14

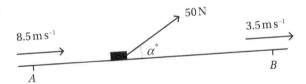

A block of mass 60 kg is pulled up a hill in the line of greatest slope by a force of magnitude 50 N acting at an angle α° above the hill. The block passes through points A and B with speeds 8.5 m s^{-1} and 3.5 m s^{-1} respectively (see diagram). The distance AB is 250 m and B is 17.5 m above the level of A. The resistance to motion of the block is 6 N. Find the value of α. **[11]**

Cambridge International AS & A Level Mathematics 9709 Paper 41 Q7 Nov 2014

15 Particle P travels along a straight line from A to B with constant acceleration $0.05 \, \text{m s}^{-2}$. Its speed at A is $2 \, \text{m s}^{-1}$ and its speed at B is $5 \, \text{m s}^{-1}$.

   i Find the time taken for P to travel from A to B, and find also the distance AB.   **[3]**

Particle Q also travels along the same straight line from A to B, starting from rest at A. At time t s after leaving A, the speed of Q is $kt^3 \, \text{m s}^{-1}$, where k is a constant. Q takes the same time to travel from A to B as P does.

   ii Find the value of k and find Q's speed at B.   **[5]**

Cambridge International AS & A Level Mathematics 9709 Paper 41 Q5 Nov 2012

16 Particle A of mass $3\,m$ kg and particle B of mass m kg are at rest on a smooth horizontal plane. Initially, A is 2 metres away from B. A accelerates uniformly towards B and collides with B directly after 1.5 seconds.

   i Calculate the acceleration of A.

After the collision, the speed of A is halved and both A and B are moving in the same direction.

   ii Find the speed of B immediately after the collision.

17

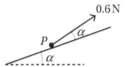

A particle P of mass 0.5 kg rests on a rough plane inclined at angle α to the horizontal, where $\sin \alpha = 0.28$. A force of magnitude 0.6 N, acting upwards on P at angle α from a line of greatest slope of the plane, is just sufficient to prevent P sliding down the plane (see diagram). Find

   i the normal component of the contact force on P,   **[2]**

   ii the frictional component of the contact force on P,   **[3]**

   iii the coefficient of friction between P and the plane.   **[2]**

Cambridge International AS & A Level Mathematics 9709 Paper 41 Q3 Nov 2012

18 Two particles, P and Q of masses $7\,m$ and $3\,m$ respectively, are moving in opposite directions along the same horizontal line. The particles collide directly and the speeds of P and Q immediately before the collision are $5 \, \text{m s}^{-1}$ and $4 \, \text{m s}^{-1}$ respectively. Immediately after the collision, the direction of Q is reversed and its speed is $10 \, \text{m s}^{-1}$. Find the speed of P immediately after the collision and its direction of motion.

19 Particles P and Q are moving in a straight line on a rough horizontal plane. The frictional forces are the only horizontal forces acting on the particles.

   i Find the deceleration of each of the particles given that the coefficient of friction between P and the plane is 0.2, and between Q and the plane is 0.25.   **[2]**

At a certain instant, P passes through the point A and Q passes through the point B. The distance AB is 5 m. The velocities of P and Q at A and B are $8 \, \text{m s}^{-1}$ and $3 \, \text{m s}^{-1}$, respectively, both in the direction AB.

   ii Find the speeds of P and Q immediately before they collide.   **[5]**

Cambridge International AS & A Level Mathematics 9709 Paper 41 Q4 Nov 2013

20 The top of a cliff is 40 metres above the level of the sea. A man in a boat, close to the bottom of the cliff, is in difficulty and fires a distress signal vertically upwards from sea level. Find

   i the speed of projection of the signal given that it reaches a height of 5 m above the top of the cliff, [2]

   ii the length of time for which the signal is above the level of the top of the cliff. [2]

The man fires another distress signal vertically upwards from sea level. This signal is above the level of the top of the cliff for $\sqrt{(17)}$ s.

   iii Find the speed of projection of the second signal. [3]

Cambridge International AS & A Level Mathematics 9709 Paper 41 Q3 June 2013

21 Particles A of mass 0.65 kg and B of mass 0.35 kg are attached to the ends of a light inextensible string which passes over a fixed smooth pulley. B is held at rest with the string taut and both of its straight parts vertical. The system is released from rest and the particles move vertically. Find the tension in the string and the magnitude of the resultant force exerted on the pulley by the string. [5]

Cambridge International AS & A Level Mathematics 9709 Paper 41 Q2 Nov 2011

22 A particle P starts at the point O and travels in a straight line. At time t seconds after leaving O the velocity of P is v m s^{-1}, where $v = 0.75t^2 - 0.0625t^3$. Find

   i the positive value of t for which the acceleration is zero, [3]

   ii the distance travelled by P before it changes its direction of motion. [5]

Cambridge International AS & A Level Mathematics 9709 Paper 41 Q4 June 2012

23 A car of mass 1250 kg travels from the bottom to the top of a straight hill of length 600 m, which is inclined at an angle of 2.5° to the horizontal. The resistance to motion of the car is constant and equal to 400 N. The work done by the driving force is 450 kJ. The speed of the car at the bottom of the hill is 30 m s^{-1}. Find the speed of the car at the top of the hill. [5]

Cambridge International AS & A Level Mathematics 9709 Paper 41 Q2 June 2013

24 A car of mass 1200 kg moves in a straight line along horizontal ground. The resistance to motion of the car is constant and has magnitude 960 N. The car's engine works at a rate of 17 280 W.

   i Calculate the acceleration of the car at an instant when its speed is 12 m s^{-1}. [3]

The car passes through the points A and B. While the car is moving between A and B it has constant speed V m s^{-1}.

   ii Show that $V = 18$. [2]

At the instant that the car reaches B the engine is switched off and subsequently provides no energy. The car continues along the straight line until it comes to rest at the point C. The time taken for the car to travel from A to C is 52.5 s.

   iii Find the distance AC. [5]

Cambridge International AS & A Level Mathematics 9709 Paper 41 Q7 Nov 2012

Particles P and Q have masses $2\,m$ kg and $5\,m$ kg respectively. The particles are moving towards each other on a smooth horizontal plane and they collide directly. Immediately before the collision, the speeds of P and Q are $4\,u$ m s^{-1} and ku m s^{-1} respectively. Immediately after the collision, the direction of motion of both particles is reversed and their speeds are halved. Find the value of k.

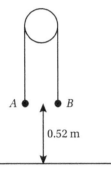

26

Particles A and B, of masses 0.3 kg and 0.7 kg respectively, are attached to the ends of a light inextensible string. The string passes over a fixed smooth pulley. A is held at rest and B hangs freely, with both straight parts of the string vertical and both particles at a height of 0.52 m above the floor (see diagram). A is released and both particles start to move.

i Find the tension in the string. [4]

When both particles are moving with speed 1.6 m s^{-1} the string breaks.

ii Find the time taken, from the instant that the string breaks, for A to reach the floor. [5]

Cambridge International AS & A Level Mathematics 9709 Paper 41 Q6 Nov 2013

 A particle P starts from a point O and moves along a straight line. P's velocity t s after leaving O is v m s^{-1}, where
$$v = 0.16t^{\frac{3}{2}} - 0.016t^2.$$

P comes to rest instantaneously at the point A.

i Verify that the value of t when P is at A is 100. [1]

ii Find the maximum speed of P in the interval $0 < t < 100$. [4]

iii Find the distance OA. [3]

iv Find the value of t when P passes through O on returning from A. [2]

Cambridge International AS & A Level Mathematics 9709 Paper 41 Q7 Nov 2011

28

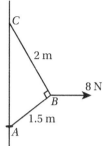

A small ring of mass 0.2 kg is threaded on a fixed vertical rod. The end *A* of a light inextensible string is attached to the ring. The other end *C* of the string is attached to a fixed point of the rod above *A*. A horizontal force of magnitude 8 N is applied to the point *B* of the string, where *AB* = 1.5 m and *BC* = 2 m. The system is in equilibrium with the string taut and *AB* at right angles to *BC* (see diagram).

i Find the tension in the part *AB* of the string and the tension in the part *BC* of the string. **[5]**

The equilibrium is limiting with the ring on the point of sliding up the rod.

ii Find the coefficient of friction between the ring and the rod. **[5]**

Cambridge International AS & A Level Mathematics 9709 Paper 41 Q7 June 2012

29

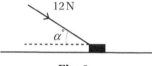 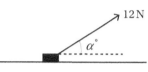

| Fig. 1 | Fig. 2 |

A block of mass 2 kg is at rest on a horizontal floor. The coefficient of friction between the block and the floor is μ. A force of magnitude 12 N acts on the block at an angle α to the horizontal, where $\tan\alpha = \frac{3}{4}$. When the applied force acts downwards as in Fig. 1 the block remains at rest.

i Show that $\mu \geqslant \frac{6}{17}$. **[5]**

When the applied force acts upwards as in Fig. 2 the block slides along the floor.

ii Find another inequality for μ. **[3]**

Cambridge International AS & A Level Mathematics 9709 Paper 41 Q5 Nov 2011

30 A particle *P* starts from rest at a point *O* and moves in a straight line. *P* has acceleration $0.6t$ m s^{-2} at time *t* seconds after leaving *O*, until $t = 10$.

i Find the velocity and displacement from *O* of *P* when $t = 10$. **[5]**

After $t = 10$, *P* has acceleration $-0.4t$ m s^{-2} until it comes to rest at a point *A*.

ii Find the distance *OA*. **[7]**

Cambridge International AS & A Level Mathematics 9709 Paper 41 Q7 Nov 2013

Extension questions

1 A snooker cue with a velocity of $0.8\,\text{m}\,\text{s}^{-1}$ hits a white ball of mass $0.17\,\text{kg}$, causing it to collide with a red ball of mass $0.16\,\text{kg}$, which is stationary at a distance of $24\,\text{cm}$ away. After the collision, the white ball is at rest. Given that momentum is conserved during the collisions and the coefficient of friction between the snooker table and the balls is 0.1, calculate the total distance travelled by the red ball before it comes to rest. **[10]**

2 A particle accelerates from rest such that $a = \sin 2t + 2$. Calculate the velocity and displacement of the particle after ten seconds. **[6]**

3 A particle is pulled across a smooth horizontal table by four forces at bearings of $035°$, $107°$, $203°$ and $311°$. The magnitude of each force in newtons is equal to its bearing. Calculate the magnitude and direction of the resultant force. **[5]**

4 40 individual forces, F_1 to F_{40}, are applied to a particle P.

F_1 has magnitude $1\,\text{N}$ towards the north.

F_2 has magnitude $2\,\text{N}$ towards the east.

F_3 has magnitude $3\,\text{N}$ towards the south.

F_4 has magnitude $4\,\text{N}$ towards the west.

F_5 has magnitude $5\,\text{N}$ towards the north.

F_6 has magnitude $6\,\text{N}$ towards the east.

...

F_{40} has magnitude $40\,\text{N}$ towards the west.

Find the magnitude and the direction of the resultant of the 40 forces.

5 $X\text{N} \longleftarrow\!\bullet\!\longrightarrow Y\text{N}$

A particle P is pulled to the left by a force of $X\,\text{N}$ and to the right by a force of $Y\,\text{N}$. The magnitude of the forces X and Y is determined by rolling a fair six-sided dice twice. The first roll determines the magnitude of X and the second roll determines the magnitude of Y.

i What is the probability that P accelerates to the right?

ii What is the maximum probability that P moves either left or right with a constant velocity?

6 A ball A is dropped from a height $30\,\text{m}$ above the ground and falls vertically downwards. One second later, a second ball B is projected vertically upwards with an initial velocity of $u\,\text{m}\,\text{s}^{-1}$.

i Given that the balls are at the same height when ball A has fallen two thirds of the distance to the ground, find the value of u.

ii Given that the balls do not collide, calculate the exact amount of time between the two balls hitting the ground for the first time.

7 In this question, you may use the result

$$\sum_{r=1}^{n} r^2 = \frac{n}{6}(n+1)(2n+1).$$

A series of 19 particles are released from rest at 1 second intervals. Each particle accelerates at a constant rate of $4\,\text{m}\,\text{s}^{-2}$. Find the cumulative distance travelled by the 19 particles after 19 seconds.

8 A snowball has an initial mass of 2 kg and begins to accelerate down a hill with an initial velocity of $1\,\mathrm{m\,s^{-1}}$. It is given that

$$\frac{\mathrm{d}v}{\mathrm{d}t} = 2t^2 - 2$$

and

$$\frac{\mathrm{d}m}{\mathrm{d}t} = 3t^2 - 2.$$

Find the change in momentum for the snowball during the first 3 seconds of its motion.

9

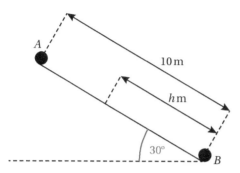

A particle A is released from rest and accelerates down a smooth slope inclined at an angle of 30° to the horizontal. At the same time that A is released from rest, a particle B is projected up the slope with an initial speed of $8\,\mathrm{m\,s^{-1}}$. The particles collide when B has moved a distance h m up the slope. Find the value of h and the time taken for the particles to collide.

10 A particle P of mass $3u$ kg is projected vertically upwards with an initial velocity of $u\,\mathrm{m\,s^{-1}}$. After T seconds, the velocity of P is halved and P is $\frac{u}{16}$ metres above its initial position. Calculate the loss in kinetic energy after T seconds and find the value of T.

11 Initially, particles P and Q, each of mass 2 kg, are a distance of $5u$ metres apart. P is projected towards Q with an initial velocity of $u\,\mathrm{m\,s^{-1}}$ and it undergoes a constant resistive force of 2 N. Simultaneously, Q is projected vertically upwards with a velocity of $v\,\mathrm{m\,s^{-1}}$.

i Find an expression in terms of u for the minimum value of v such that P passes the original position of Q before Q returns to the ground.

ii Find the range of possible values for u for which the solution to **part a** is valid.

12 A particle P of mass m kg accelerates uniformly as it moves in a straight line from A to B in 5 seconds. Between A and B, the change in momentum is $10\,\mathrm{kg\,m\,s^{-1}}$ and the change in kinetic energy is 25 joules. Find the distance between A and B.

GLOSSARY

acceleration The rate of change of velocity.

coalesce If two bodies coalesce then they join upon impact and move together as one mass.

coefficient of friction The coefficient of friction is represented by μ in the formula $F \leqslant \mu R$, where F is the frictional force and R is the normal reaction. For a smooth surface $\mu = 0$.

collide One or more moving objects striking each other.

component When resolving a force into two parts, each of the two parts is called a component.

conservation of energy A fundamental principle stating that energy cannot be created or destroyed, only converted from one form to another.

conservation of linear momentum When two objects moving along the same line of motion collide, and there are no external forces acting on them, the total momentum before the collision is equal to the total momentum after the collision.

constant of integration When a function is differentiated to find a gradient function, any constant terms will become equal to 0. Consequently when any gradient function is integrated to identify the curve function an arbitrary constant is introduced. This constant is known as the constant of integration.

deceleration As for acceleration, the rate of change of velocity, but specifically when the body is slowing down.

differentiation The process of finding an expression for a rate of change or gradient. Differentiating displacement leads to velocity. Differentiating velocity leads to acceleration.

displacement The distance moved in a certain direction from an object's starting point. Displacement is a vector quantity as it has magnitude and direction.

displacement–time graph Graph of displacement against and time. The gradient at any point on the graph is the instantaneous velocity.

distance How far an object travels.

energy A measure of the ability of a body to do work.

equilibrium If a system is in equilibrium, then the resultant force in any direction is zero.

force When an object experiences a push or pull it is being acted on by a force. Weight, tension, thrust and friction are examples of forces. Force is a vector quantity and is measured in newtons (N).

friction The force that opposes motion along a rough surface.

gravity The force of attraction between two bodies. On Earth this force, gravity, causes an acceleration, g, of approximately $10 \, \text{m s}^{-2}$.

integration The opposite process to differentiation. Integrating acceleration leads to velocity. Integrating velocity leads to displacement.

joule The SI unit used for energy and work done.

kinetic energy The energy a body has as a result of its motion. The kinetic energy, KE, of an object is found using the formula $\text{KE} = \frac{1}{2}mv^2$, where m is the mass (kg)

of the body and v (m s^{-1}) is the velocity of the body.

limiting equilibrium An object is in equilibrium on a rough surface when the force due to friction, F, and the normal reaction, R, satisfy $F \leqslant \mu R$, where μ is the coefficient of friction. When the object is on the point of slipping, it is in limiting equilibrium, and $F = \mu R$.

magnitude The size (or length) of a vector.

mass The mass of a body is the amount of matter it contains.

momentum The tendency of a moving object to continue moving. The momentum, p, of an object is found using the formula $p = mv$, where m is the mass of the object and v is its velocity.

Newton's first law A particle remains at rest or continues to move with a constant velocity in a straight line unless acted upon by an external force.

Newton's second law The force, F, acting upon a particle is proportional to the particle's mass, m, and acceleration, a. This results in the formula $F = ma$.

Newton's third law Every action has an equal and opposite reaction.

normal reaction If an object is in contact with a surface, the object will experience a reaction force perpendicular to the surface. This is the normal reaction, R.

potential energy The energy a body has as a result of its height. The potential energy of an object, PE, is found using the formula $\text{PE} = mgh$, where m is the mass (kg) of the object, g is the acceleration due to gravity $(10 \, \text{m s}^{-2})$ and h is height (m).

power The rate at which a force does work. Power = Fv, where F is the force (N) and v is the velocity (ms^{-1}).

pulley A wheel over which a string passes, often for the purpose of raising an object.

resolving The process of separating a single force into two parts is called resolving.

resultant force The overall force is called the resultant. For example, the resultant of two forces P and Q is the single force that could take the place of these two forces.

rough If a surface is rough, it will produce a resistive frictional force.

scalar A quantity with magnitude but no direction.

SI units An internationally recognised system of units.

smooth If a surface is smooth, it will produce no resistive frictional force.

speed The magnitude of the velocity.

tension The force in a string when it is taut and being pulled by an object.

thrust A force that pushes an object.

vector A quantity that has magnitude (size) and direction.

velocity The rate of change of displacement. Velocity is a vector quantity as it has magnitude and direction.

velocity–time graph Graph of velocity against time. The gradient at any point on the graph is the instantaneous acceleration and the area under the graph is the displacement.

watt The SI unit of power.

weight The weight of an object is a force caused by the gravitational acceleration experienced by the object. It acts vertically downwards.

work done The work done against a force of F N for a distance of d m is given by the formula $WD = Fd$.

Modelling word	Assumption
Light	The object has no mass
Smooth	There is no friction
Rough	There is friction
Inextensible / Inelastic	The object cannot be stretched or squashed
Uniform	The same throughout (for example, uniform velocity)
Particle	A single point representing an object
Rigid	The object cannot bend
String	A line with no thickness
Rod	A rigid straight line with no thickness

INDEX